04-WEM-1

FACTORS FOR SUCCESS IN DEVELOPING USE ATTAINABILITY ANALYSES

by:

Paul L. Freedman, P.E., BCEE
LimnoTech

Thomas Dupuis, P.E.
CH2M HILL

2007

WERF

The Water Environment Research Foundation, a not-for-profit organization, funds and manages water quality research for its subscribers through a diverse public-private partnership between municipal utilities, corporations, academia, industry, and the federal government. WERF subscribers include municipal and regional water and wastewater utilities, industrial corporations, environmental engineering firms, and others that share a commitment to cost-effective water quality solutions. WERF is dedicated to advancing science and technology addressing water quality issues as they impact water resources, the atmosphere, the lands, and quality of life.

For more information, contact:
Water Environment Research Foundation
635 Slaters Lane, Suite 300
Alexandria, VA 22314-1177
Tel: (703) 684-2470
Fax: (703) 299-0742
www.werf.org
werf@werf.org

This report was co-published by the following organizations. For non-subscriber sales information, contact:

IWA Publishing
Alliance House, 12 Caxton Street
London SW1H 0QS, United Kingdom
Tel: +44 (0) 20 7654 5500
Fax: +44 (0) 20 7654 5555
www.iwapublishing.com
publications@iwap.co.uk

Library of Congress Catalog Card Number: 2007923525
IWAP ISBN: 978-1-84339-786-1

LimnoTech, CH2M HILL

The research on which this report is based was developed, in part, by the United States Environmental Protection Agency (EPA) through Cooperative Agreement No. CR-83155901-1 with the Water Environment Research Foundation (WERF). However, the views expressed in this document are solely those of LimnoTech and CH2M HILL and neither EPA nor WERF endorses any products or commercial services mentioned in this publication. This report is a publication of WERF, not EPA. Funds awarded under the Cooperative Agreement cited above were not used for editorial services, reproduction, printing, or distribution.

ACKNOWLEDGMENTS

Report Preparation

Principal Investigators:
Paul Freedman, P.E., BCEE,
LimnoTech

Thomas Dupuis, P.E.
CH2M HILL

Project Manager:
Hans Holmberg, P.E., BCEE
LimnoTech

Project Team:
Michele Koehler
Adrienne Nemura, P.E.
Virginia Breidenbach, P.E.
Wendy Larson
Ruth Blum
LimnoTech

Sherrill Doran
Linda Macpherson
CH2M HILL

Patricia McGovern, P.E.
Patricia McGovern Engineers

Lori Terry
Foster Pepper PLLC

Project Subcommittee

Fred Andes, *Barnes & Thornburg*
Thomas Gardner, *U.S. Environmental Protection Agency*
Tom Grovhoug, *Larry Walker Associates*
Keith Linn, Northeast *Ohio Regional Sewer District, PSC Research Council Liaison*
Jodi Perras, *Perras & Associates*
Lynn Sisk, *Alabama Dept. of Environmental Management*

Water Environment Research Foundation Staff

Director of Research: Daniel M. Woltering, Ph.D.
Program Director: Margaret Stewart

ABSTRACT AND BENEFITS

Abstract:

A Use Attainability Analysis (UAA) is a process to review and potentially modify a waterbody's designated uses, as embodied in the state or tribal water quality standards. A UAA is considered when a designated use assigned to a waterbody is not an "existing use" and attaining the designated use is not feasible because of one or more natural or human-caused conditions. Completing a successful UAA can be challenging and it is confounded by a range of technical, regulatory, and institutional considerations that must be addressed. This research identifies, highlights, and analyzes the major challenges faced in the UAA process, and reports on the common lessons learned from how these challenges were addressed. Researchers conducted a detailed analysis of more than 50 UAAs, selected from an examination of over 200 UAA efforts with diverse geographic, topical, and waterbody representation. On the basis of this review, 29 findings were developed, and were validated by close examination of case studies. Special attention was focused on three common and challenging situations where UAAs are being considered nationwide: wet weather impacts; urban settings; and effluent-dependent or -dominated streams. The findings, or factors for success, focused on five broad categories of criteria: scientific and technical understanding; legal and regulatory requirements; financial needs and impacts; public awareness and involvement; and regulatory agency preparedness. These findings were transformed into recommendations for activities or approaches that would be supportive of UAA efforts. Case studies illustrating the findings and recommendations are provided along with summaries of selected state programs. The overall result of this research is a user's guide to UAAs. This guidance is designed to inform stakeholders of critical factors in UAAs and related information needs. The findings and recommendations are intended to help those involved make well-informed, well-supported, and well-documented decisions throughout the UAA process.

Benefits:

- Provides a user's guide for anyone considering conducting a UAA, already participating in a UAA, and states that need to evaluate or prepare guidance related to UAAs.
- Describes factors for success to help users conduct UAAs more efficiently and effectively.
- Illustrates critical factors for UAA success through case studies.
- Includes six detailed case studies with detailed illustrations of the key challenges faced in the UAA process, and how best to address those challenges.
- Describes state protocols for UAA frameworks, which will assist states that are reviewing UAAs or interested in developing UAA guidance, as well those undertaking individual UAA efforts.

Keywords: Use Attainability Analysis, UAA, designated uses, water quality criteria, water quality standards, user's guide.

TABLE OF CONTENTS

Acknowledgments ... iii
Abstract and Benefits ... iv
List of Tables ... vii
List of Figures ... viii
List of Acronyms ... ix
Executive Summary ... ES-1

1.0 Introduction ... 1-1
- 1.1 Project Objectives ... 1-1
- 1.2 Background ... 1-1
- 1.3 Overview of the User's Guide ... 1-5
- 1.4 Other Resources ... 1-6

2.0 Research Methods and Criteria ... 2-1
- 2.1 Identifying Case Studies ... 2-1
- 2.2 Defining Success in Context of the Process and Outcome ... 2-2
- 2.3 Description of Research Criteria and Questions ... 2-4
- 2.4 General Questions to Clarify UAA Status and Background ... 2-4
- 2.5 Scientific and Technical Understanding ... 2-6
- 2.6 Financial Needs and Impacts ... 2-8
- 2.7 Legal and Regulatory Requirements ... 2-9
- 2.8 Political Support ... 2-9
- 2.9 Public Awareness and Involvement ... 2-10
- 2.10 Regulatory Agency Preparedness ... 2-11
- 2.11 Developing Findings and Recommendations from the Case Studies ... 2-12

3.0 Research Findings ... 3-1
- 3.1 Scientific and Technical Findings ... 3-1
- 3.2 Financial Needs and Impacts ... 3-6
- 3.3 Legal and Regulatory Requirements ... 3-10
- 3.4 Regulatory Agency Preparedness ... 3-13
- 3.5 Public Awareness and Involvement ... 3-17

4.0 Guidance on Characterizing and Conducting Successful UAAs ... 4-1
- 4.1 Introduction ... 4-1
- 4.2 Translating Findings into Recommendations ... 4-2
- 4.3 Identifying the Stakeholders ... 4-7
- 4.4 Integrating Recommendations into the UAA Process ... 4-8
- 4.5 Hypothetical Examples of Decision Pathways ... 4-18

5.0 Highlighted Case Studies 5-1
5.1 Boston Harbor Combined Sewer Overflows 5-1
5.2 California Engineered Flood Channels 5-4
5.3 Chesapeake Bay 5-8
5.4 Cuyahoga River Ship Channel 5-13
5.5 Spokane River 5-16
5.6 Lower French Gulch and Blue River, Colorado 5-20

6.0 Statewide UAA Protocols 6-1
6.1 Recreational UAA Protocols 6-3
6.2 Aquatic Life UAAs 6-6
6.3 Legislation to Address Uses During Wet Weather 6-9

7.0 Identified Needs 7-1
7.1 UAA Database 7-1
7.2 Guidance Manuals 7-1
7.3 Technical Guidance on Standards 7-2
7.4 Dissemination of Findings and Tools 7-2

Appendix A: Resource Panel Contact Information A-1
Appendix B: Case Study Inventory B-1
Appendix C: Use Attainability Analysis Resources C-1

References R-1

LIST OF TABLES

ES-1 UAA Integrated Recommendations ES-4
2-1 Characteristics of the 54 UAA Efforts That Were Examined in Detail through Application of Research Criteria 2-3
4-1 Crosswalk between Findings (Criteria for Success; Chapter 3.0) and Recommendations (Guidance; Chapter 4.0), Scientific and Technical Understanding... 4-3
4-2 Crosswalk between Findings (Criteria for Success; Chapter 3.0) and Recommendations (Guidance; Chapter 4.0), Legal and Regulatory Requirements. 4-4
4-3 Crosswalk between Findings (Criteria for Success; Chapter 3.0) and Recommendations (Guidance; Chapter 4.0), Financial Needs and Impacts........ 4-5
4-4 Crosswalk between Findings (Criteria for Success; Chapter 3.0) and Recommendations (Guidance; Chapter 4.0), Public Awareness and Involvement. 4-6
4-5 Crosswalk between Findings (Criteria for Success; Chapter 3.0) and Recommendations (Guidance; Chapter 4.0), Regulatory Agency Preparedness............ 4-7
4-6 UAA Integrated Recommendations 4-10
6-1 Overview of Statewide UAA Protocols and State Legislation to Address Wet Weather Issues Discussed in Chapter 6.0........ 6-2
6-2 Maine's Water Quality Classification........ 6-8

LIST OF FIGURES

ES-1 Evaluation Criteria for Factors Leading to Success ES-2
4-1 Integrated Framework for a Successful UAA 4-1
5-1 Map of Boston Harbor Area 5-3
5-2 Map of Greater Los Angeles Region 5-7
5-3 Chesapeake Bay UAA Study Area 5-10
5-4 Cuyahoga River Study Area 5-14
5-5 Spokane River Study Area 5-18
5-6 Lower French Gulch and Blue River, Colorado, Watershed 5-21

LIST OF ACRONYMS

CDPHE (DPHE)	Colorado Department of Public Health and Environment
CE-QUAL-W2	Corps of Engineers, Water Quality, 2-Dimensional Model
CERCLA	Comprehensive Environmental Response, Compensation, and Liability Act
CFR	*Code of Federal Regulations*
CSO	combined sewer overflow
DO	dissolved oxygen
F-	financial needs and impact
FERC	Federal Energy Regulatory Commission
KDHE	Kansas Department of Health and Environment
L-	legal and regulatory
LTCP	long-term control plan
MA DEP	Massachusetts Department of Environmental Protection
MDNR	Missouri Department of Natural Resources
MEPA	Massachusetts Environmental Protection Act
MWRA	Massachusetts Water Resources Authority
NACWA	National Association of Clean Water Agencies
NOAA	National Oceanographic Atmospheric Agency
NPDES	National Pollutant Discharge Elimination System
PA-	public awareness
POTW	publicly owned treatment works
R-	regulatory agency preparedness
RAP	remedial action plan
RWQCB	Regional Water Quality Control Board
S&T-	scientific and technical
TMDL	total maximum daily load
UAA	Use Attainability Analysis
U.S. EPA	U.S. Environmental Protection Agency
USGS	United States Geological Survey
WBCR	whole body contact recreation
WERF	Water Environment Research Foundation

Executive Summary

State and tribal water quality standards include a process to review and potentially modify a waterbody's designated uses, known as a Use Attainability Analysis (UAA). A UAA is considered when a designated use assigned to a waterbody is not an "existing use" and attaining the designated use is not feasible because of one or more natural or human-caused conditions]. At this time, there is no detailed comprehensive UAA guidance from U.S. EPA, which makes undertaking and completing a successful UAA challenging. A range of technical, regulatory, and institutional considerations must be addressed in the UAA process. These challenges prompted the need for this research project.

The objective of this research was to identify factors for success in the UAA process, and to then develop guidance for users to apply these factors. Primary emphasis during the research was on collecting and comprehensively evaluating case studies, building on the successes already achieved, and creating an awareness of the obstacles that have led to failures. This report summarizes observational findings gained from reviewing UAA case studies, which were transformed into recommendations for how to better conduct a UAA. Case studies supporting the findings are presented, as well as summaries of effective state protocols.

ES.1 Background

UAAs are often undertaken when problems arise complying with regulatory programs such as total maximum daily load (TMDL), National Pollutant Discharge Elimination System (NPDES), or combined sewer overflow (CSO) long-term control plans (LTCPs). Discharges into highly urbanized watersheds and into effluent-dominated waterbodies, as well as wet-weather discharges, may further drive the need for a UAA. Although there are many reasons for conducting a UAA, regulations require a UAA anytime a designated use is removed or less stringent water quality criteria are applied.

The federal Clean Water Act places the responsibility on individual states to adopt water quality standards for their lakes, rivers, and streams. The Clean Water Act has the goal that, wherever attainable, the standards provide for the protection and propagation of fish, shellfish, and wildlife and provide for recreation in and on the water. These goals are often referred to as "fishable/swimmable". Many states adopted standards that were generalized and not site-specific in meeting these goals. If the designated uses for the water are not correctly assigned, then regulatory control efforts may not be protecting an appropriate use, and may be committing public resources to protect a use that is not feasible to attain. However, the law and its implementation in federal regulations implicitly recognized that the adoption of standards is an adaptive process with some changes expected. A UAA can provide the justification for a change to water quality standards notwithstanding the need to meet the fishable/swimmable goals where attainable.

Federal regulations in 40 *CFR* 131.10(g) set forth factors that may be used to justify removing or modifying designated uses, including (in summary): the presence of naturally occurring pollutants; low flow or physical restrictions that prevent uses; human caused conditions or sources of pollution and hydrologic modifications that are not feasible to reverse; and proposed control strategies that might result in substantial and widespread economic and social impact. Any one of these factors can justify a change in use, except that a designated use cannot be removed if it is an "existing" use, or a use that can be attained by implementing technology-based effluent limits required under the Clean Water Act.

The outcome of a UAA can be any one of the following: no change; a revised use that provides greater protection; removal of a use; a refined use; partial uses or a temporary suspension or modification of a use; or site-specific criteria to protect a use. A common misconception by the public is that the intent of a UAA is to permit increased pollutant loadings and will automatically cause a degradation of the quality or current uses of the waterbody. In reality, a UAA refines the standards to more accurately characterize site-specific conditions.

ES.2 Research Approach

Factors supporting the successes in UAA efforts were identified in this research through detailed examination of 54 UAA case studies. These case studies were culled from a candidate inventory of over 200 UAAs by focusing on those that provided good illustrations of success factors and/or provided diverse geographic, topical and waterbody representation. The 54 UAA efforts were examined through application of a questionnaire that provided a comprehensive framework of information on the UAAs in the context of five categories of criteria for the factors. Based on this information, various hypotheses regarding factors supporting success were posed. These hypotheses were subsequently confirmed (or refuted) through validation by case studies. The findings were then transformed into recommendations for activities or approaches that would be supportive of UAA efforts.

The evaluation criteria were developed to identify critical factors in UAA success. The categorical criteria used for this research fell into five groups, as shown in Figure ES-1. For each criterion, a list of research questions was developed to help the research team investigate the process for each UAA case study.

Figure ES-1. Evaluation Criteria for Factors Leading to Success.

The findings are a combination of qualitative observational patterns, conventional wisdom and professional experience, but not statistically validated correlations. Unfortunately, the sample set was too limited to provide for a rigorous validation of findings, as each UAA is unique and replication of procedures rarely occurred. The findings are illustrated by case study examples.

The development of recommendations for navigating the UAA process was a relatively straightforward process. Each finding led to a specific recommendation, which often was merely to follow or avoid activities in each factor, depending on the situation. Some of the recommendations were not so easily drafted but the link to the findings is still apparent.

ES.3 Findings and Recommendations

Twenty-nine findings were derived from the detailed evaluation of the case studies. These observations or findings provided significant insights as to the factors contributing to success. However, to be more clear and explicit, specific recommendations were developed in response to each finding that the researchers felt were practices that would promote success. These findings and recommendations were then integrated into an overall process as summarized in Table ES-1.

The tendency to think about and focus on each recommendation independently is enticing because a UAA is a complex undertaking. However, to be successful, integrated recommendations need to be considered in each phase of the UAA process. For purposes of discussing recommendations in an integrated plan, the UAA process has been divided into four phases, getting started, conducting the UAA, involving stakeholders, and resolution.

The findings and recommendations were integrated into this framework to give the reader of this guide a sense of how they fit into the overall UAA process.

Getting Started

Most recommendations are best initiated at the beginning of the UAA process. Investing time and resources at the start of the UAA process leads to a more efficient and effective UAA. Generally, investments in the following three activities provide the foundation for a successful UAA effort: agreeing on who should lead and be involved in the UAA; agreeing on a strategy to satisfactorily address potential issues for all relevant agencies and stakeholders; and defining a decision pathway and the required data needs.

Conducting the UAA

Once a UAA team, strategy, and decision pathway have been defined, two elements contribute to the successful conduct of a UAA. First, sound science with good documentation and record keeping is critical to support the required technical analysis. This is necessary to not only convince concerned stakeholders but also to set a solid foundation for a successful outcome. The second important element is combining technical and economic analyses to define what is "feasible." Neither technical nor economic analyses alone may be sufficient to justify a use change.

Table ES-1. UAA Integrated Recommendations.

Getting Started
1. Identify and involve all affected stakeholders
a. Establish decision process ground rules for the role and expectations of the stakeholder process (S&T-3, PA-1)
b. Discuss whether there is a need for impartial third-party facilitation or mediation (PA-2, PA-4)
c. Identify partners and collaborate on the identification of common issues—this can help to reduce costs and leverage scarce resources and build support (L-5, PA-5)
d. Use a collaborative process to promote the use and/or development of state guidance even if from other states (F-5, R-1)
e. Involve all regulators early in the process, learn their issues and make certain that there is a willingness to comply with rulemaking schedules and guidance (S&T-1, L-1; R-1)
2. Develop a work plan and strategy for the UAA
a. Work with all relevant agencies and stakeholders to understand all the issues and regulatory constraints (PA-2, PA-3, L-3)
b. Develop a mutually agreed upon strategy that addresses all issues and identified constraints (S&T-1, PA-2, R-3, R-4)
c. Integrate the UAA with other regulatory programs and agencies (S&T-6, L-3, R-3)
d. Engage U.S. EPA Regional staff early on (R-4)
3. Develop a decision pathway
a. Define the decision pathway and relevant data needs (S&T-3)
b. Identify and define applicable factors and critical regulatory terms (S&T-2, L-1, L-2)
c. Work with agencies to define the scope, by making use of any available guidance (R-2, S&T-2)
d. Scope the data and modeling needs upfront to know what the UAA will cost (F-4, S&T-2)
e. Use models to help answer attainability questions in complex systems (S&T-4)
f. Identify and involve downstream users (L-6, PA-2)
Conducting the UAA
1. Work to gain approval
a. Integrate sound science and with a thoughtful, inclusive stakeholder process (S&T-1, S&T-2, L-4, PA-1)
b. Keep good documentation (L-2)
c. Request mediation if litigation seems a likely result of conflict (PA-1, PA-3, PA-6)
2. Develop thorough and appropriate economic analyses (as needed)
a. Consider including economic analysis (e.g., cost-benefit or cost-effectiveness) to help stakeholders understand what may be "feasible" for technical factor UAAs (S&T-5, PA-5, PA-6)
b. Conduct an "affordability" evaluation if Factor 6 is a legal basis for the UAA; consider other indicators of affordability beyond U.S. EPA economic guidance (F-1)
c. Additional economic guidance for UAAs is needed (F-2)
d. Consider whether putting an economic value on the use is needed (F-3)
Involving Stakeholders
1. Fully engage the public
a. Do more public involvement than the minimum (PA-1)
b. Use face-to-face communication (PA-4)
c. Inform stakeholders on the purpose and potential outcome of UAA (PA-3)
d. Identify the economic investments at stake and educate the public of the consequences (PA-6)
2. Keep your decision-makers in the loop
a. Communicate the issues to real decision-makers (R-3, R-4, PA-5)
b. Define economic impacts for elected officials (PA-6, S&T-5)
Resolution
1. Be open to alternative regulatory outcomes (R-5)

Involving Stakeholders

Beyond the technical and legal elements of the UAA, the research also has shown that effective stakeholder involvement and consensus are key elements to success. Two basic recommendations were identified: 1) UAA projects need to fully engage the public and all relevant stakeholders, most effectively achieved through face-to-face communication; and 2) politicians can provide valuable leadership even though they are not conducting the UAA.

Resolution

The last phase of the integrated guidance is resolution. A successful UAA outcome can be defined broadly. Alternative regulatory solutions can result from consideration of a UAA. For example, a UAA may lead to widespread public and political recognition that current designated uses are appropriate. Revisiting the stakeholders' objectives and needs is a good idea because it represents being open to an alternative regulatory solution. Examples of alternative solutions are many but could include a temporary variance, a long-term compliance schedule with milestones for reconsideration, a use refinement versus a use removal, site-specific water quality criteria, a long-term consent order, or additional evaluations to support regulatory analyses or requirements.

The research made it clear there is no one path to a successful UAA. However, the findings from the research included many good lessons learned that can be applied by anyone considering or pursuing a UAA.

ES.4 Case Studies and State Guidance

Six detailed case studies were selected from the 54 UAAs researched. These case studies illustrate the key challenges faced in the UAA process and highlight how best to address those challenges, consistent with the findings and recommendations presented in this report. The case studies were selected to represent three common and challenging situations where UAAs are being considered nationwide: wet-weather impacts; urban settings; and effluent dependent or dominated streams.

Five statewide UAA protocols addressing both recreational and aquatic life uses were also identified in the research. In general, the UAA protocols described in this report were developed in response to statewide use designation issues. The protocols serve to streamline UAA efforts, in some instances where numerous UAAs have to be conducted in a short time period, and may save costs. In addition, the guide presents an example of a state with a tiered aquatic life use system, and three examples of states that have developed legislation to address designated uses during wet weather. The examples presented include the following: recreational UAA protocols in Colorado, Kansas, and Missouri; aquatic life UAAs examples in Ohio and Oklahoma; tiered aquatic life uses in Maine; and state legislation to address wet-weather uses in Indiana, Maine, and Massachusetts.

CHAPTER 1.0

INTRODUCTION

A process to review and potentially modify a waterbody's designated uses, as embodied in the state or tribal water quality standards is known as a Use Attainability Analysis (UAA). A UAA is considered when a designated use assigned to a waterbody is not an "existing use" [see 40 *CFR* 131.3(e)] and attaining the designated use is not feasible because of one or more natural or human-caused conditions [see 40 *CFR* 131.10(g)]. The process of undertaking, let alone completing, a successful UAA has proven to be challenging. One key reason for the difficulty in conducting successful UAAs is that there is no detailed comprehensive UAA guidance from U.S. EPA. The UAA process is confounded by the range of technical, regulatory, and institutional considerations that must be addressed in conducting a successful UAA. These challenges prompted the need for this research project.

This report is a resource for those who are considering conducting a UAA for a waterbody within their watershed, or those who are already participating in a UAA. It is also intended to be a resource to states that need to evaluate or prepare UAA-related guidance. The guide summarizes the findings of a detailed review and analyzes 54 UAA efforts from all across the country, selected from an examination of more than 200 UAA efforts.

1.1 Project Objectives

This research was to identify factors for "success" in the UAA process, and then develop guidance for users to apply these factors. The emphasis was placed on assisting stakeholders to become quickly aware of critical factors and information requirements to make well-informed decisions related to their involvement. The guidance in this document is grounded in actual experiences with UAA efforts, with primary emphasis on collecting and comprehensively evaluating case studies, building on the successes already achieved and creating an awareness of the obstacles that have led to failures.

1.2 Background

This section provides background information on UAAs, including a general description of the process and the regulatory basis for UAAs, a discussion of the need for conducting UAAs to determine appropriate water quality standards, and a summary of potential outcomes of a UAA.

1.2.1 The UAA Process

A UAA is a process to support the review and potential modification of a waterbody's designated uses, or an associated water quality criterion. In 40 *CFR* 131.3(g) a UAA is defined as: "a structured scientific assessment of the factors affecting the attainment of the use which may include physical, chemical, biological, and economic factors as described in 131.10(g)." Although a UAA may be pursued for many reasons, most often UAAs are considered in connection with one or more of the following activities:

- Permit issuance or renewal within the National Pollutant Discharge Elimination System program (NPDES);
- Listing of a waterbody as "impaired" under Section 303(d) of the Clean Water Act, which may require a total maximum daily load (TMDL);
- Draft or completed TMDLs that specify a reduction in the amount of pollutants that can be discharged into a waterbody; and
- CSO LTCP, which trigger the need to reevaluate the designated use(s) of a waterbody.

Discharges into highly urbanized watersheds and into effluent-dominated waterbodies may further drive the need for a UAA. Although reasons may differ for conducting a UAA, regulations require a UAA anytime a designated use is removed or less stringent water quality criteria are applied.

With the passage of federal water quality legislation (e.g., federal Water Pollution Act Amendments of 1972 and the Clean Water Act in 1977), the goal of the nation's leaders was to clean up waterways so that they were "fishable and swimmable" and to restore and maintain the "chemical, physical, and biological integrity of the Nation's waters" (Section 101a of Pl92-500). Often overlooked, however, is the fact that the water quality goals of the Clean Water Act also included the words "where attainable." Adopting water quality standards was one of the first steps to achieving the fishable and swimmable goal. Water quality standards consist, in part, of a specific desired use appropriate to the waterbody (designated use), and narrative and/or numeric criteria to protect the designated use. A UAA is a regulatory process that allows water quality standards to be reviewed and revised when the uses upon which the standards are based are neither existing nor attainable.

The Clean Water Act places the responsibility on individual states to adopt water quality standards, and on U.S. Environmental Protection Agency (U.S. EPA) to approve or disapprove them. However, the law implicitly recognized that the adoption of standards would have to be an adaptive process with some changes expected, hence the creation of the UAA process. In hindsight, this was important because many states adopted standards that were generalized and not site-specific and/or were viewed more as a vision or goal. In either case, many of these standards might not be suitable for regulatory purposes and might warrant a UAA.

Federal regulations allow states flexibility to remove or modify water quality standards through the UAA process. States may remove a designated use that is not an existing use, as defined in 131.3, or establish subcategories of a use if the state can demonstrate that attaining the designated use is not feasible [40 *CFR* 131.10(g)]. The regulations set forth the factors that may be used to modify designated uses:

- Factor 1 - Naturally occurring pollutant concentrations prevent the attainment of the use; or
- Factor 2 - Natural, ephemeral, intermittent or low-flow conditions or water levels prevent the attainment of the use, unless these conditions may be compensated for by a sufficient volume of effluent discharge without violating state conservation requirements to enable uses to be met; or
- Factor 3 - Human-caused conditions or sources of pollution prevent the attainment of the use and cannot be remedied, or would cause more environmental damage to correct than to leave in place; or

- Factor 4 - Dams, diversions or other types of hydrologic modifications preclude the attainment of the use, and it is not feasible to restore the waterbody to its original condition or to operate such modification in a way that would result in the attainment of the use; or
- Factor 5 - Physical conditions related to the natural features of the waterbody preclude attainment of aquatic life protection uses; or
- Factor 6 - Controls more stringent than those required by sections 301(b) and 306 of the Act would result in substantial and widespread economic and social impact.

Any one of these factors can justify a use change, as is made clear by the use of the word "or" following each UAA factor. In addition to these permissive factors, certain constraints are also placed on UAAs. A designated use cannot be removed if it is an "existing" use, a use that was "actually attained in the waterbody on or after November 28, 1975" [40 *CFR* 131.3(e)]. Also, a use cannot be removed if the use can be attained by implementing technology-based effluent limits required under sections 301(b) and 306 of the Clean Water Act for point sources and cost-effective and reasonable best management practices for nonpoint source control [40 *CFR* 131.10(d)].

Regulations do not state who conducts or leads the analysis. In practice, a UAA can be initiated and led by states, stakeholders, tribes, or U.S. EPA. Usually, more than one of these groups is involved in the process; however, before a use can be changed the state must adopt the use change and that change must be approved by U.S. EPA. Many states, agencies, and other parties who have pursued a UAA have found the process difficult. Part of the reason is often associated with a lack of understanding of the specific technical, biological, and economic information necessary to meet one or more of the UAA regulatory factors set forth in 40 *CFR* 131.10(g)(1)-(6). Public perception that a UAA will degrade a waterbody, as well as the expense associated with some UAA studies, has also complicated UAA efforts. As a result, UAAs are not being undertaken for many waterbodies that might warrant consideration of a standards change.

1.2.2 The Need for Appropriate Water Quality Standards

In a recent U.S. EPA memorandum (King, 2006) the Director of the Office of Science and Technology strongly supported the need for "credible and defensible UAAs" as the vehicle to "getting the uses right." Director Ephraim King underlined several important aspects to improve the UAA process, many of which are highlighted in the findings and recommendations of this report. A significant observation was the importance of the UAA process to national clean water programs, even if the specific steps are still vague and undefined.

In 1972 the Clean Water Act required water quality standards. This led to a rush by states in the late 1970s to adopt water quality standards to be eligible for federal infrastructure grant programs. In addition, states lacked the resources to finely tune uses based on ecoregional or seasonal conditions, or to assign uses to every waterbody. As a result, many waterbodies were broadly designated, sometimes resulting in designated uses that were not existing uses or could never be attained.

Likewise, general nonspecific uses were often assigned to larger waterbodies and then applied to their tributaries. Since the late 1970s, after significant progress was made in implementing minimum technology-based wastewater treatment levels, there have been regulatory trends that make assigning the appropriate designated use more important.

One of these regulatory trends has been the shift from technology-based effluent limitations to receiving stream water quality-based effluent limitations. This means that instead of focusing effluent limitations on what is technically achievable or practical with respect to wastewater treatment, the focus is increasingly on what is required in terms of effluent quality to support designated uses in the receiving water. In NPDES permits, water-quality based effluent limits are based on meeting water quality objectives that support the designated uses of the receiving water. If the designated uses for the water are not correctly assigned, then permit holders may not be protecting an appropriate use, and/or may be committing public resources to protect a use that is not feasible to attain.

Another trend has been the emergence of the TMDL program as the centerpiece for water quality restoration efforts. The TMDL process consists of listing waters not attaining standards and developing point and nonpoint source loading allocations designed to achieve these standards. The TMDL process is driven by the designated uses and associated water quality criteria. If the designated use assignment is not appropriate, the TMDL may be designed to protect a use that is not feasible to attain, a result that is not intended under the Clean Water Act. U.S. EPA now recommends better integration of the UAA process and TMDL development (King, 2006).

Nonpoint and wet-weather sources create unique issues in the context of meeting water quality standards because of the difficulty of tracking these sources and the expenses associated with controlling them. Given the challenge of developing affordable and available controls for nonpoint and wet-weather sources, it is essential that water quality standards be based on existing and attainable designated uses. More specifically for wet-weather sources, the current emphasis on developing LTCPs for CSOs to meet water quality standards is another trend that has placed more importance on the UAA process. U.S. EPA's CSO Control Policy outlines the need to review and revise, as appropriate, the applicable water quality standards. The price tag for implementing LTCPs for large cities is often a billion dollars or more, therefore, having realistic attainable standards as the regulatory target is critical.

Use designation is also a significant consideration for effluent-dependent waterbodies. In recent years, growth and expansion into arid areas of the country has resulted in increased discharges of treated effluent into intermittent and low-flow streams. Under these conditions, the discharge dominates the physical conditions in the streams, which can lead to a number of issues surrounding the appropriate assignment of a designated use. Many effluent-dependent waterbody dischargers are now dealing with the repercussions by having to meet water standards at end-of-pipe. Another issue is the question of "healthy" aquatic populations when aquatic life exists solely dependent on the discharge. In many waterbodies, the year-round effluent creates a perennial stream. This in turn changes the type of aquatic habitat compared to the original intermittent and ephemeral stream. What then is the appropriate designated use and associated water quality criteria?

In all of these cases, the UAA process can be helpful in determining the appropriate use.

1.2.3 Outcomes of the UAA Process

A common misconception by the public is that a UAA will result in a degradation of the quality of the waterbody. In reality, a UAA is undertaken to refine the standards so that they more accurately characterize the site-specific conditions of the waterbody. It does not permit an increase in pollutant loading or allow for degraded water quality. According to U.S. EPA (King, 2006), a credible UAA can result in a change in the designated use in either direction. The result of a UAA may refine the use description or change the use to restrict or expand the uses. In all cases, the UAA must support a designated use that reflects existing and attainable uses and the corresponding criteria to be protective of those uses.

One of the challenges faced in complying with current water quality standards is the assumption that every waterbody can achieve the assigned designated use, such as primary contact recreation, at all times. Obviously, this does not always hold true. For example, during large storm events there are short periods of time when a waterbody may not be swimmable for reasons not related to water quality. However, various federal and state policies do not currently consider this reality outside of the UAA process.

Through the UAA process a new designated use can be assigned and/or revised water quality criteria can be established. In fact, there are many possible outcomes of the UAA process, many of which do not result in an explicit change from one designated use to another.

One outcome of the UAA process may be a partial use designation. A partial use designation allows the use to change during an event driven occurrence, such as storm events, or seasonal changes, such as low stream flows limiting fish passage. Another outcome may be the refinement or addition of new, specific uses or subcategories of use created based on ecoregional areas or other site-specific conditions. It is important to note that with these types of use designations, new water quality criteria may also be established that protect the new or refined use. Therefore, the result of these UAAs will include more detailed narrative descriptions of the desired use as well as more finely stratified measurable criteria.

Another outcome of the UAA process may be a variance. A variance is a temporary modification of the water quality standard that may be appropriate for the short-term, typically renewed every three years. A variance may be used until better data or information is available, a new technology is developed, costs for existing technologies decrease, or other modifications can be made so that the use can be met. The same six factors that allow a modification of a designated use also define reasons for which a variance can be implemented and approved.

Ultimately, the outcome of a UAA should result in identifying the appropriate designated use for the waterbody to protect public health and the environment.

1.3 Overview of the User's Guide

Included in this guide is a summary of the findings of a detailed review and analysis of 54 UAA efforts from all across the country, selectively culled from an examination of over 200 UAA efforts. The detailed reviews were selected to specifically identify, highlight and analyze the major challenges faced in the UAA process and report on the common lessons learned from how these challenges were addressed. The objective of this research was to identify factors for success and provide guidance to users on how to make these factors work for them. The findings, developed from the review of case studies, focus on the factors most critical to success or failure in the following areas of the UAA process:

1. Scientific and Technical Understanding
2. Financial Needs and Impacts
3. Legal and Regulatory Requirements
4. Public Awareness and Involvement
5. Regulatory Agency Preparedness

Important elements related to effluent-dependent dischargers, wet-weather impacts, and highly urbanized areas have been highlighted. This guidance provides the findings and recommendations of this investigation and provides the reader with criteria for success in completing a UAA. Thus, the findings and recommendations in this user's guide are heavily influenced by research and analysis of numerous real-world UAA case studies.

1.4 Other Resources

The research team highly recommends that those interested in UAAs also obtain and use another guidance handbook, "Collaborative Water Quality Solutions: Exploring Use Attainability Analyses, developed cooperatively by Water Environment Research Foundation and the National Association of Clean Water Agencies (WERF & NACWA, 2005). The UAA Handbook provides extensive information regarding procedural, policy, legal, economic, and technical aspects of UAAs; history of and future prospects for UAAs; stakeholder involvement tools and processes tailored for UAAs; and other special considerations. This handbook, along with this user's guide, provides the practitioner with complete and comprehensive information and guidance on navigating the UAA process. Additional resources are provided in Appendix C.

CHAPTER 2.0

RESEARCH METHODS AND CRITERIA

Researchers identified factors supporting successful UAAs through detailed examination of 54 UAA case studies that provided good illustrations of success factors and/or also provided diverse geographic, topical, and waterbody representation. On the basis of questionnaire responses and interviews on each of five success factors, various hypotheses were posed regarding factors supporting success. Those hypotheses that were subsequently confirmed through validation by case studies are presented as “findings.” The findings were then transformed into recommendations for activities or approaches that would be supportive of UAA efforts.

2.1 Identifying Case Studies

A candidate list of UAA efforts was compiled by conducting a thorough search of UAA efforts across the country. The research team obtained case study candidates from WERF, NACWA, the WERF Project Subcommittee experts and resource panel member recommendations based on their nationwide project experience. The research team also contacted water quality standards coordinators in the U.S. EPA Regional offices and selected state coordinators seeking more information on UAA efforts in each region and state.

More than 200 case studies were compiled and classified in a matrix for tracking and sorting. Each UAA effort was initially classified based on information such as waterbody type, designated use, parameter of concern, U.S. EPA region, state, and status of the effort. Many of these studies were either unsuitable examples or similar to other UAAs in characteristics and process. Therefore, the list was pared down to 54 case studies for detailed examination by the research team, several of which represented groups of similar UAAs within one state.

The final list of UAA case studies was selected based on those that would be most useful in identifying factors for success. Studies that were duplicative, largely incomplete, poor illustrations or had poor documentation were excluded. Emphasis was placed on finding studies involving urban, wet-weather, and effluent dependent settings. However, all UAAs were initially considered because they had potential to illustrate factors for success even if in the context of a problem, such as mining, that was not within the emphasis of this research. The studies were also selected to ensure broad and balanced representation based on geography, U.S. EPA region, designated use, pollutant of concern, waterbody type, and factors used for UAA justification.

UAA efforts identified for research represented all 10 U.S. EPA regions, 30 states, the District of Columbia, and one US Territory. These UAAs were distributed among eight designated uses assessed, 18 parameters of concern or pollutant source types, and seven waterbody types. The case studies are summarized in Table 2-1; Appendix B provides more details on each case study.

2.2 Defining Success in Context of the Process and Outcome

This research raises the fundamental question of how to define or recognize a successful UAA process. Here, success was viewed not only in the context of an approved UAA, but also in context of the process and how stakeholders viewed it as a tool in resolving the relevant water quality challenges.

Most UAAs are undertaken because of a regulatory problem that is tied to water quality standards. Therefore success is often viewed in the context of resolving that problem. However, in this project other criteria for selecting case studies were considered, including the following:

- The UAA process led to designated use changes that were adopted by the state and approved by U.S. EPA, and which resolved the regulatory issue related to water quality standards (such as NPDES permit, LTCP, TMDL, etc.).
- The UAA process led to an alternative regulatory solution (such as variances, site-specific criteria, revised TMDL, alternative NPDES permit requirements, etc.).
- The UAA process validated the current designated uses as appropriate or perhaps recognized that additional uses were appropriate, which then led to broad public and political support for needed improvements in water quality.
- The UAA process led to incomplete or unresolved issues but the process provided insight into the UAA methodology.

Although only the first category represents a completed and successful change in designated use, the next two were also viewed as successes for selection of case studies. This was because the UAA process led to either stakeholder acceptance of the designated use and/or a regulatory solution of the note worthy problem. The last category might be viewed as a failure, but functionally these "failures" often provided useful insights into the UAA process. For this reason, a few incomplete or unresolved UAAs were also included in the case studies.

Different stakeholder groups or individuals involved in any UAA process are likely to have different perspectives on success, and it is critical that these perspectives be understood and respected throughout the UAA process. Because the focus in this project was on the process, the project team also included case studies that are either incomplete or unresolved. Consequently, this research project did not attempt to formally classify any of the UAAs that were investigated as being successful or not successful, or to quantify how many or what percentage were "successful." Rather, all the available case studies were analyzed in the context of different elements in the UAA process according to the various criteria described in this chapter. Findings and recommendations were then developed that do not rely on labeling any particular UAA an overall "success." Rather, the *focus was on the process for each UAA effort.*

Table 2-1. Characteristics of the 54 UAA Efforts That Were Examined In Detail through Application of Research Criteria.

Note: Total numbers or percentages for each key characteristic do not total 54 or 100% because multiple descriptions may apply for a UAA effort.

Key Characteristic	Description	Total (#)	Total (%)
Setting	Effluent Dominated	27	50%
	Wet-weather impacted	16	30%
	Urban	18	33%
	Other	23	43%
Pollutant Source Category	Agricultural runoff	19	35%
	CSOs	9	17%
	Industry	14	26%
	Natural background	14	26%
	POTWs	22	41%
	Urban stormwater	16	30%
	Mining	6	11%
	Other	9	17%
UAA Factor - 131.10(g)	1	14	26%
	2	22	41%
	3	27	50%
	4	15	28%
	5	16	30%
	6	11	20%
	Undetermined	3	6%
Designated Use Assessed	Aquatic life	36	67%
	Recreation	23	43%
	Drinking water	11	20%
	Other	3	6%
Parameters of Concern	Bacteria	14	26%
	Metals	6	11%
	Nutrients	14	26%
	Organic hydrocarbons	0	0%
	Pesticides/Herbicides	0	0%
	Solids	5	10%
	Temperature	10	19%
	Ammonia	4	7%
	BOD	5	9%
	Chloride	3	6%
	Dissolved oxygen	12	22%
	Other	19	35%
Status	Completed and adopted	29	54%
	Completed and rejected	8	15%
	Considered but never developed	4	7%
	Started but then abandoned	1	2%
	Ongoing	13	24%
	Not started but under consideration	6	11%
	Completed and under U.S. EPA review	2	4%
Outcome	Use upgrade	6	11%
	Use removal/revision	28	52%
	No change in use	15	28%
	Other	25	51%

2.3 Description of Research Criteria and Questions

Before initiating research into past and ongoing experiences with UAAs, evaluation criteria were developed to guide the research. These evaluation criteria ensured consistent and comprehensive assessments of the UAAs identified for research that did not rely on the outcome of the UAA effort. The evaluation criteria were developed to advance the primary purpose of the project, which was to produce a user's guide to help stakeholders become quickly aware of critical factors and information requirements to make well informed decisions related to UAAs. These criteria allowed the research team to analyze each UAA effort in the context of its various component processes and attributes. The evaluation criteria were used to identify lessons learned and ways to improve the development and acceptance of UAAs in the future. The categorical criteria proposed for this research fall into five groups:

1. Scientific and Technical Understanding (S&T)
2. Financial Needs and Impacts (F)
3. Legal and Regulatory Requirements (L)
4. Public Awareness and Involvement (PA)
5. Regulatory Agency Preparedness (R)

For each criterion, a list of research questions was developed to help the research team investigate the process for each UAA case study. These research questions were divided into two groupings: (1) upper-level hierarchy questions; and (2) detailed questions. The upper-level hierarchy questions were designed to ensure that the research team acquired the essential information associated with each criterion for each identified UAA. The detailed questions were used by the research team to understand the complexity of the UAA process and to delve deeper into each UAA on an as-needed basis.

The research criteria and questions were initially drafted by the research team. They were then provided to both the Expert Panel and Project Subcommittee for review and comment. Useful revisions were made to these criteria as a result of this input. The team also tested the evaluation criteria to two UAA cases: Fenholloway River in Florida and Chicago Waterways. This process aided the team in revising the evaluation criteria and the style of the interview when necessary. The research team then applied the revised evaluation criteria to the UAA analysis.

The criteria for each category and the research questions that were used to assess the criteria are presented in the following sections. The detailed questions are included in their entirety in this chapter because they will be helpful to users of this guidance for understanding the potential complexity of the UAA process and the questions that form each UAA process.

2.4 General Questions to Clarify UAA Status and Background

The evaluation of each UAA began with the following general questions that provide a basic understanding of the status, background, and setting of the UAA.

- Who is taking the lead in developing the UAA?
- When was the UAA started and when is it expected to be completed?
- What is the status of the UAA process?
  - completed and approved
  - completed and rejected

    - considered but never developed
    - started but then abandoned
    - on-going
    - under consideration
    - under U.S. EPA review
- What are the waterbodies included in the UAA effort?
- In which state(s) is the UAA effort being conducted?
- What type(s) of waterbody(ies)?
    - Estuary
    - Large river
    - Small river
    - Inland lake
    - Great lake
    - Coastal
    - Intermittent stream
    - Other (describe)
- What designated uses are being assessed?
    - Aquatic life
    - Recreation
    - Other (describe)
- What is the setting in which the UAA is taking place?
    - Urban
    - Wet-weather impacted
    - Effluent dominated
- What are the primary pollutant sources?
    - Publicly owned treatment works (POTWs)
    - CSOs
    - Urban stormwater
    - Industry
    - Agricultural runoff
    - Natural background
    - Other (describe)
- What are the water quality parameters of concern?
- Bacteria
    - Metals
    - Nutrients
    - Organic hydrocarbons
    - Pesticides/herbicides
    - Solids
    - Temperature
    - Other (describe)
- What is the UAA condition being pursued [40 *CFR* 131.10(g)(1)-(6)]?
    - (1) Naturally occurring pollutant concentrations prevent the attainment of the use
    - (2) Natural, ephemeral, intermittent, or low-flow conditions or water levels prevent the attainment of the use

- (3) Human-caused conditions or sources of pollution prevent the attainment of the use
- (4) Dams, diversions, or other types of hydrologic modifications preclude the attainment of the use
- (5) Physical conditions related to the natural features of the waterbody preclude attainment of aquatic life protection uses
- (6) Controls more stringent than those required by sections 301(b) and 306 of the Act would result in substantial and widespread economic and social impact
- Undetermined

- Contact information
- Documentation

2.5 Scientific and Technical Understanding

The research criteria and associated questions are described in detail below.

Criterion: Scientific and technical factors were sufficiently defined and understood to support the need for preparation of UAA (i.e., designated uses cannot be met).

Upper-level Hierarchy of Research Questions: Which of the five technical factors in 40 *CFR* 131.10(g)(1)-(5) were used for the UAA (recognizing that only one of the factors is necessary for a UAA decision)? Did the decision-makers at the state regulatory agency define what information was needed to make a decision before the UAA? Did they find the analysis and data adequate or was additional study required? What type of data/information led to a successful outcome? How did the UAA fit into a TMDL process (if at all)? Did attainment of a use, or attainment of a criterion, drive the development of the UAA? What contributed to an unsuccessful outcome? How did scientific and technical understanding of the waterbody affect the successful or unsuccessful outcome of the UAA?

Detailed Research Questions

- What water quality information (physical, chemical, and biological parameters) was used to support development of the UAA? How current was it? Did decision makers and stakeholders feel the data was sufficient or was additional monitoring required? How did the UAA fit into a TMDL process (if at all)? Did attainment of a use, or attainment of a criterion, drive the development of the UAA?
- How were watershed boundaries and/or subbasin relationships defined and were they accounted for through watershed plans and/or models?
- Were quantitative or qualitative descriptions of pollutant sources associated with perceived impairment well known or identified (e.g., mass balance of point and nonpoint sources)? If so, how were such descriptions used?
- How were current and future land uses and the application of best management practices within the watershed considered for their impact on attainment/nonattainment?
- Were water quality models used to define current and future water quality conditions? What specifically was modeled? Were modeling assumptions and outcomes accepted by stakeholders?
- How were current designated uses and criteria established? For example, have they been on the books for decades without any reconsideration historically? Were uses

developed in recognition of reasonable and cost-effective best management practices and point source controls and the 40 *CFR* 131.10(g) factors. Were site-specific conditions taken into account for establishing criteria? If not, were generic definitions of uses and supporting criteria adopted? Were these same considerations part of the UAA process?

- How were downstream water quality standards evaluated and considered?
- How was attainment/nonattainment of designated uses determined (e.g., direct measures such as beach closures, shellfish bed closures, recreation surveys, or biological surveys or indirect measures such as nonattainment of water quality criteria)? If indirect, how was nonattainment of criteria assessed and did that necessarily indicate nonattainment of uses?
- How were existing uses defined or identified (including as far back as November 1975 to meet the regulatory definition of an "existing" use)?
- Were naturally occurring pollutants, flow conditions, human-caused conditions, hydrologic modifications, and/or natural physical conditions sufficiently understood to support the feasibility assessment for the five respective technical factors in 40 *CFR* 131.10(g)(1)-(5), in relation to both existing and attainable uses (the more specific questions below can be used to answer this more general question, recognizing that one of these factors must be satisfied to remove a use that is not an existing use, or to establish a subcategory of use with less stringent water quality criteria)? Were any of the below issues critical to the success or failure of the UAA? If applicable:
  - Did the UAA compare conditions in reference waterbodies or watersheds to the waterbody in question [40 *CFR* 131.10(g)(1)-(5)]?
  - How were aquatic life and habitat conditions evaluated [40 *CFR* 131.10(g)(2)-(5)]?
  - How were naturally occurring sources vs. human caused sources of pollutants evaluated and quantified [40 *CFR* 131.10(g)(1)]?
  - How were natural, ephemeral, intermittent, or low-flow conditions or water levels understood and evaluated [40 *CFR* 131.10(g)(2)]?
  - How were human caused conditions such as watershed changes (e.g., urbanization identified and assessed for their contribution towards attainment/nonattainment [40 *CFR* 131.10(g)(3)]?
  - How was the ability to remedy human caused conditions assessed to sufficiently understand the feasibility of attaining designated uses [40 *CFR* 131.10(g)(3)]?
  - How were hydrologic modifications (dams, channelization, diversions, etc.) identified and assessed for their contribution towards attainment/nonattainment [40 *CFR* 131.10(g)(4)]?
  - How was the feasibility of restoration and/or operational modifications associated with hydromodifications addressed in relation to attainment of uses [40 *CFR* 131.10(g)(4)]?
  - If applicable, how was information on the natural physical conditions (see categories below) evaluated [40 *CFR* 131.10(g)(5)]?
    - hydrology
    - geophysical/morphology (e.g., substrate, riffle/pools, etc.)
    - atmospheric
    - surface and groundwater interaction

2.6 Financial Needs and Impacts

Criterion: Appropriate financial information was used to define substantial and widespread economic and social impact (if used as a factor for supporting UAA) or "feasibility" of attaining current water quality standards [40 *CFR* 131.10(g)(6)].

Upper-level Hierarchy of Research Questions: Were costs and other financial considerations primary issues in the UAA process? Were they included in the UAA process, and if so how was it done and what guidance was used? Was it a formal Factor 6 analysis? Was cost-effectiveness or cost-benefit (e.g., knee-of-the –curve) analysis included or the main analysis?

Detailed Research Questions:

- Did the financial analysis in the UAA differentiate between and identify cost/benefit, cost-effectiveness, financial capability, and affordability concepts, before initiating the analysis? If so, which method of analysis was used?
- Were there U.S. EPA and/or state guidelines to follow? If so, were they followed? If not, was some other guidance or recipe book followed (e.g., another state's or U.S. EPA region's guidance)?
- Was demographic information up to date in relation to growth patterns and infrastructure planning?
- Was household income distribution information available and current? If so, how was this information used?
- How were economic and social impacts on sensitive low-income households evaluated?
- Were the full spectrum of other governmental services/utility costs and financial conditions well defined, and were community resource limitations considered in light of achieving the highest overall quality of life in a community? If so, how was that done? Or was the analysis limited only to a small subset of services/costs such as tertiary treatment or wet-weather controls, for example in relation to a TMDL process?
- Were water quality standards changes considered for any waterbodies in minority or economically disadvantaged neighborhoods? If so, was environmental justice considered?

Criterion: The cost of preparing and adopting a water quality standards change based on a UAA was fully developed at an "order of magnitude" level of understanding during the scoping process.

Upper-level Hierarchy of Research Question: Did the key proponents of the UAA process have a clear understanding of the scope and cost that would be involved in completing the UAA? Were costs estimated at an "order of magnitude" level ($10K vs. $100K vs. $1M) or to a more detailed level?

Detailed Research Questions:

- Were data collection and/or compilation costs considered?
- Were modeling costs considered?
- Were coordination requirements (organization, stakeholder involvement, and regulatory consultation) considered? This includes the ultimate need for U.S. EPA approval of state water quality standards.
- Was the schedule realistic and was it consistent with the budget?

- Were scientific, technical, planning, legal and administrative costs considered?
- Were management costs considered?

Criterion: Sufficient information was available to understand the value of the waterbody use under current conditions, at the current designated use, and at the use supported by a UAA.

Upper-level Hierarchy of Research Questions: Were economic valuations of the waterbody use (e.g., beach visits, commercial or recreational fishing revenues) developed for various levels of water quality conditions? If so, how were these used?

2.7 Legal and Regulatory Requirements

Criterion: The requirements under state and federal law to develop a defensible UAA and proposed revisions to water quality standards (site specific or subcategories) were defined in the UAA scope of work.

Upper-level Hierarchy of Research Questions: What were the legal requirements and obstacles, and how were they satisfied (or not) during the UAA process?

Detailed Research Questions:

- Were existing consent decrees or other enforceable agreements/understandings considered? If so, how?
- Were other regulatory processes such as the Endangered Species Act, Federal Energy Regulatory Commission relicensing or state water right laws considered? If so, how?
- Were downstream water quality and water right requirements considered? If so, how?
- Were jurisdictional issues such as interstate waters or Tribal Government standards considered? If so, how?
- What legal procedures were followed and why, e.g., was it a rule-making? What was the legal determination in this case? What specific state and/or federal action was used to adopt the UAA?
- Were there legal challenges by nongovernmental organizations or others?

2.8 Political Support

Later in the study, these Political Support criteria were integrated with the criteria on Public Awareness and Involvement because the scope of the findings in this area was narrow and closely tied to political interest in public support.

Criterion: The need to prepare a UAA created sufficient urgency to gain political priority and leadership (e.g., willingness to publicly support the UAA process).

Upper-level Hierarchy of Research Questions: Was there a willingness on the part of the political leadership in the community or geographic area to support the UAA process? If there was political opposition, who mounted in and in what form? What were the factors that provided the urgency or willingness, or lack thereof?

Detailed Research Questions:

- Were potential economic impacts/benefits understood?
- Were potential environmental and public health impacts/benefits understood, including consideration of the extent of current and potential future uses?

- Was environmental justice considered?
- Were interstate water issues considered?
- Were regulatory conflicts considered (e.g., interagency goal conflicts)?
- Were public interests evaluated and understood?
- If there was political opposition, who mounted it and in what form?

Criterion: Leadership roles were defined and accepted (local, state, federal, tribal).

Upper-level Hierarchy of Research Questions: Were leadership roles defined and accepted? Who were the players and what were their roles?

Detailed Research Questions:

- Who was involved in the process? Should others have been more involved? Should some have been less involved?
- Who was the lead decision-maker? Was the decision-maker clearly identified?
- Who were the individuals within each entity assigned/committed to the UAA? What roles did they play in their organizations?
- How did political support (or lack thereof, or outright opposition) affect the outcome of the UAA?

2.9 Public Awareness and Involvement

Criterion: Community members had sufficient site-specific knowledge of the need for a UAA and the process for developing a UAA.

Upper-level Hierarchy of Research Questions: How were community members informed and educated about the need for a UAA and what the possible outcomes would be?

Detailed Research Questions:

- How was the public educated on water quality issues, such as watersheds, pollution sources, and potential impacts on public health and the environment?
- How was the public educated about the regulatory drivers, such as the Clean Water Act and state water quality standards (uses and criteria), NPDES, and CSO requirements? What were the obstacles to public understanding, if any?
- How was it communicated to community members that a UAA would not result in further degradation of the waterbody?
- Were community members informed of the costs to comply with current designated uses and the benefits to be gained?

Criterion: Community members informed the UAA process.

Upper-level Hierarchy of Research Questions: How did the community/public provide feedback or input to the UAA process?

Detailed Research Questions:

- Did the public have a high level of trust with current custodians (dischargers and agencies) of waterbody?
- Were community members surveyed on the past, current, and desired future uses of the waterbody?
- Was the public comfortable with the way in which current expenditures on water quality protection were being used? How do you know?

- Were community members surveyed on their willingness to pay for improved water quality? If yes, how?
- Was the linkage (or lack thereof) between water quality standards and actions required to attain these standards explained to the community (e.g., higher sewer rates don't necessarily result in improved water quality)?

Criterion: Community members supported the UAA.

Upper-level Hierarchy of Research Questions: Did the community/public ultimately support the results and outcomes of the UAA?

Detailed Research Questions:

- Did the public support the refinement of designated uses through a UAA? Why or why not?
- Were there appeals or legal challenges of the UAA decision? Why or why not?
- Did the issue become controversial (i.e., news media coverage, legislative inquiries, involvement of elected officials)? How were the controversial issues resolved?
- How did public support or opposition affect the outcome of the UAA?

2.10 Regulatory Agency Preparedness

Criterion: Water quality agencies were prepared to support development of the UAA.

Upper-level Hierarchy of Research Questions: Was the decision-making process understood by proponents of the UAA in terms of who the decision-makers were, were they involved from the beginning, and were they authorized to make decisions?

Detailed Research Questions:

- Were guidelines and procedures in place to guide review and approval procedures?
- Did management and staff recognize the need and urgency for a UAA? Did they actively support or oppose the UAA?
- Did management and staff see the UAA as a useful tool to integrate water quality programs?
- Did staff have the technical knowledge and strengths to support review and evaluation of UAA?
- What state or federal agency staff were assigned and committed to the UAA? Did they include staff responsible for permitting, water quality assessment, and water quality standards? How was upper management involved?
- Did inter-agency relationships support the integration of UAA concept of existing and attainable uses into water quality standards and resource management goals? For example, did fisheries management agencies and water quality agencies have common goals for the uses of the waterbodies in question (e.g., warm water vs. cold water goals, protection of native vs. introduced species, protection of endangered species, public safety hazards vs. recreation use designations, etc.)? If not, how did this affect the outcome of the UAA?
- How did regulatory agency preparedness affect the outcome of the UAA?

2.11 Developing Findings and Recommendations from the Case Studies

The information gathered from interviews using the above detailed questions was compiled in a database. This information was analyzed to identify findings that support success in the UAA process. Findings that were identified were translated into a specific recommendation. These recommendations were then combined into an integrated discussion of recommendations for the overall UAA process and divided into four phases: getting started; conducting the UAA; involving stakeholders; and resolution. In addition, six case studies were identified as highly illustrative of the recommendations and these are described in detail in Chapter 5.0.

The information for the questionnaire was compiled and sorted to evaluate various hypotheses about the UAA process. Because UAAs involve a range of approaches and issues, the team felt it appropriate to examine hypotheses in the context of different subsets of the data, sorted by issue such as use type, pollutant, wet weather, or effluent dominated. Hypotheses about the UAA process were then tested by looking at the full collection of results to either validate or refute the finding. Findings were validated by common observation of the hypothesis but not rigorous statistical verification.

The findings hypotheses were drafted based on close examination of the UAA process undertaken in case studies efforts. They were also based on interviews conducted by team members, consultation with the Project Subcommittee and experience of the research team. Thus, the findings are a combination of qualitative observational patterns, conventional wisdom and professional experience, but not statistically validated correlations. Unfortunately, the sample set was too limited to provide for a rigorous validation of findings, as each UAA is a unique situation and replication of procedures rarely occurred. The findings are illustrated by case study examples.

The development of recommendations for navigating the UAA process was a relatively straightforward process. Each finding led to a specific recommendation, which often was merely to follow or avoid activities in each factor depending upon the UAA situation. Some of the recommendations were not so easily drafted but the link to the findings is still apparent.

After the initial criteria were applied to the UAA efforts and major findings and recommendations from the efforts were collated, summarized, and verified, the research team selected six UAAs to present as case studies. The research team selected two case studies from each setting to best reveal widely applicable UAA processes, demonstrate particular findings, and to reflect settings most useful to the reader. These case studies are presented in Chapter 5.0. Chapter 6.0 provides an overview of state programs that were seen to be illustrative of useful practices that support a better UAA process.

CHAPTER 3.0

RESEARCH FINDINGS

Research findings were drawn from an in-depth analysis of the case studies, proceeding first with a set of hypotheses and then examining the case studies to validate those observations. The research findings are organized according to the five criteria categories described in Chapter 2.0:

1. Scientific and Technical Understanding (S&T)
2. Financial Needs and Impacts (F)
3. Legal and Regulatory Requirements (L)
4. Public Awareness and Involvement (PA)
5. Regulatory Agency Preparedness (R)

The discussion also identifies and very briefly describes a case study example applicable to each finding. More details on each of these case studies, including contact information, can be found in Appendix B. For a few select cases, additional detail is provided in Chapters 5.0 and 6.0. The discussion in this chapter is intentionally succinct with limited discussion so that the reader can more easily gain an overview of the process, which is as important as the individual components. Further, the findings are generally quite straightforward and simple in concept, needing little commentary.

3.1 Scientific and Technical Findings

In the category of S&T findings, several observations were quite apparent, but they each related to one single concept: "getting agreement on technical scope and approach at the beginning" and maintaining that agreement. The case studies clearly illustrate that success was facilitated by early involvement among the stakeholders and agencies to agree on the needed data and analyses, the appropriate decision pathways and definition of terms, the necessary modeling approach for complex sites, the context and scope for economic analyses, and the need to integrate with other regulatory programs, especially TMDLs. These findings are described below.

3.1.1 Finding S&T-1: Agency Involvement Throughout the Process was Common in Successful UAAs

Finding: Early and continuing agency involvement throughout the UAA process is more common in more recent, successful UAAs. This includes development of or agreement on UAA guidance, early project and work plan involvement, and ongoing project review.

Discussion: This finding pertains to UAAs that are sponsored or initiated by regulated entities rather than the regulatory agencies. The review of the more recent UAAs of this type reflects the recognition that agencies are much more likely to accept the outcome of the UAA if they have been involved in scoping the project and thinking about data needs and ultimate interpretation.

Case Study Example: Dell Arkansas Power Plant UAA for Drainage Ditches

The Dell Arkansas Power Plant conducted a UAA to determine the existing and attainable uses and water quality criteria for created drainage ditches. The ditches were listed for domestic water supply but were not being used as such. At its inception, UAA stakeholders provided state and federal agency decision-makers with a detailed work plan including suggested data collection and analysis for the UAA. The agencies provided feedback, the work plan was modified, and the process began. The UAA stakeholders and the agencies discussed and agreed upon UAA protocols and decision pathways. Agencies and stakeholders were involved throughout the UAA process reviewing progress and findings. This facilitated a smooth approval process for the UAA, and the domestic water supply use was removed from the drainage ditches.

Another good example is for the Lower French Gulch and Blue River in Colorado, an area impacted by past mining activities. In this case sixteen government agencies and stakeholder groups collaborated to agree on this UAA. More details on this case are described in relation to other findings.

3.1.2 Finding S&T 2: Agencies Often Required More Data and Analysis than Anticipated

Finding: When nonagencies sponsor a UAA, agencies typically request more data and a higher level of analysis. This is because agencies generally are more comfortable reaching conclusions and making decisions based on their own data. Further, different oversight agencies often have different data thresholds.

Discussion: When UAA sponsors are regulated entities, such as NPDES permit holders, the initial reaction of the agencies may be that the goal is simply to avoid needed water pollution controls to meet current standards, particularly if the agencies are not included early and openly in the process. Such preconceived notions can result in burdensome and/or changing data requirements simply to deter completion of a UAA. In other cases, agencies may feel that a higher degree of rigor regarding data and analysis is necessary in UAAs sponsored by permit holders so that the agencies can feel comfortable adopting a UAA outcome in a public rule-making process. At times, different agencies may have different data expectations, such as U.S. EPA versus a state, as in the case study example below.

Case Study Example: Lower Boise River Tributaries UAA (ID)

The local watershed advisory group sponsored this UAA, with support from the State of Idaho. During the UAA process, the project sponsors worked closely with the State (but not U.S. EPA) to develop the UAAs and form the basis for subsequent TMDLs. However, U.S. EPA and the Idaho Department of Environmental Quality had different expectations for UAA contents. A few years following the state's formal adoption of the UAAs, U.S. EPA Region 10 requested more analysis. The additional analyses focused both on addressing limitations of existing data and the perceived inadequacy of what the highest attainable uses might be. At this time, TMDLs are in hiatus because the final uses have not been determined. The project sponsors and the State are working together to develop a work plan to address U.S. EPA's request, to explicitly define what data are required and how that data will be used if the UAA process continues.

3.1.3 Finding S&T 3: Successful UAAs Generally Had a Decision Pathway Set at the Beginning

Finding: The research found that successful UAAs generally had a decision pathway set at the beginning. This pathway was most effective when it consisted of: the definition of critical regulatory terms, for example, existing use, natural, background, feasible, nonremedial, and irreversible; careful analysis of which of the 10(g) factors were appropriate; definition of data needs and how the data would be used; and a clear understanding of how the UAA process would impact various regulatory programs. Data needs were quite variable but defined by the nature of the UAA and typically included characterizing the physical, chemical, and biological aspects of the waterbody. When historical information was available, successful UAAs strived to combine this with current information.

Discussion: This finding follows logically from S&T 1 and 2. For the decision pathway to be clearly established at the beginning, the UAA proponents must obtain early input and involvement from the regulatory agencies, including agreement, to the extent possible, on how data will be interpreted and used. This process requires considerable discipline and planning. Case studies suggest that UAA studies will better result in success if the range of outcomes from data collection and analyses are considered at the beginning and their implications to the decision pathways fully considered.

Case Study Example: Lower French Gulch and Blue River (Colorado) UAA

The Lower French Gulch and Blue River in Colorado were impacted by past mining activities that were undergoing Comprehensive Environmental Response, Compensation, and Liability Act (CERCLA) investigation and remediation efforts. This UAA included a significant multi-agency planning effort at the beginning that laid out the decision pathway and data needs. The multi-agency group (including 16 agencies and stakeholder groups) worked together to define the goal of the UAA and develop the approach. The goal was defined simply as developing technically justifiable site-specific water quality standards for the stream segments under consideration. The approach developed for the UAA can be summarized as follows:

- Characterize existing conditions in terms of aquatic life, physical habitat, stream hydrology and water column chemistry;
- Project future water quality conditions that are expected to exist after the implementation of the CERCLA water treatment plant at the Wellington-Oro Mine site;
- Use literature information and professional judgment to determine the aquatic life potential for the stream segments in the study area recognizing future water quality conditions and any existing physical, biologic or hydrologic limitations to aquatic life;
- Propose site-specific water quality standards that are protective of potential aquatic life based on the conditions after CERCLA remediation.

Following this approach, the group agreed on a UAA that resulted in revised uses and criteria. The revised standards were also used for TMDL delisting and development activities. Further information on this case study is provided in Chapter 5.0.

3.1.4 Finding S&T 4: Models Were Usually Needed in Complex Systems

Finding: Models were usually needed in complex systems, including urban areas, to answer the attainability question for aquatic life uses. The modeling generally provided an objective tool to answer hypothetical "what if" questions about various levels of pollution control resulting in attainable uses.

Discussion: Although compilation and analysis of current and historical field monitoring data can usually answer the question of what uses are currently "existing" uses, the questions regarding what uses *might* be attainable can typically not be answered with data alone. Case studies showed that for complex sites the "what if" was best answered with models. This often involved the application of watershed and water quality computer models to analyze the improvements in water quality that could occur with various scenarios or pollution control options. Scenarios often included different degrees of wastewater treatment, CSO controls, stormwater best management practices, etc. The modeling scenarios also often included different water management policies that could dramatically affect attainment of aquatic life uses such as dam operations and water diversions. Watershed and water quality models were generally also needed for the other regulatory processes that may have prompted the UAA in the first place, for example TMDLs.

Case Study Example: Spokane River UAA (Washington) (CH2M HILL, 2004) and Buck, Valley, Flint, and Lost Creeks, UAA (Alabama)

A complex model for dissolved oxygen (DO) and eutrophication (CE-QUAL-W2) was developed for the Spokane River and Long Lake Reservoir by the State of Washington for a DO TMDL. The UAA sponsors agreed to use this model to determine how progressively increasing the levels of treatment for point sources would affect improvements in DO. The modeling showed that treatment levels beyond those proposed by point sources would have minimal DO benefit, and that the current DO criteria for the reservoir were not attainable regardless of the level of treatment. These issues could not be addressed with field data alone because the levels of treatment being considered were substantially more restrictive than current treatment technologies in place. In addition, a new discharge was also being proposed. Further information on this case study is provided in Chapter 5.0.

In another UAA, Alabama Department of Environmental Management (ADEM) used models to conduct UAAs for several creeks receiving wastewater effluent and agricultural runoff. Water quality modeling was used to demonstrate what was needed to attain the seasonal limited warmwater fishery use and to demonstrate that this use could not be met in the summer, even with limit-of-technology controls at wastewater treatment plants. The receiving water models were deemed important in demonstrating the fishable classification was not attainable.

3.1.5 Finding S&T 5: Economic Analyses Coupled with Technical Analyses Were Effective in Determining Feasibility

Finding: The research found that economic analyses coupled with technical analyses were effective in determining feasibility. Successful UAAs that were based, in part, on Factor 4 (human-caused conditions that cannot be remedied) were often coupled with economic or cost/benefit analysis. Relying on a combination of technical and economic issues was an effective means of defining the limits of point source and nonpoint source control.

Discussion: It can be argued that it is physically possible for almost any human-caused conditions to be remedied if enough money is spent on the problem. As a result, several of the successful case studies approached the question of what is a feasible remedy by analyzing the marginal economic costs and marginal water quality benefit as one component of the technical UAA analyses. This did not involve direct application of 10(g) Factor 6 (socioeconomic impact analysis); however, this approach is inconsistent with U.S. EPA's previous thinking on this issue as identified below:

"Feasibility could be based on technical considerations, such as the ability to operate an impoundment in an efficient manner that does not degrade water quality, as U.S. EPA intended when it originally wrote the regulation. Alternatively, some have suggested that feasibility could be based on economic considerations or a balanced consideration of cost and technology (U.S. EPA's current thinking is that the term "feasible" in use removal criterion number 4, regarding the operation of dams, should continue to refer to technical feasibility and not to economic feasibility. Criterion number 6, not number 4, is the appropriate avenue to address economic feasibility of attaining the designated use because it establishes an appropriate test of economic infeasibleness)." (*Fed. Regist.* **1998**, *63*, 36,756).

Case Study Example: Squaw Creek (California) UAA

In this case, human caused conditions from historical copper mines in the watershed were a factor in preventing attainment of uses (Factor 4). Historic mining activity in the watershed had caused the elimination of all fish species within Squaw Creek. Over the last 25 years, mining activities have since ceased and remediation is occurring to some degree, resulting in a 95% reduction in loadings over the past 25 years and a slow improvement in the ecosystem and environmental health of the Creek. The mines are doing all they technically can do and the practical attainability of uses was under question. The regulatory agency desired to see money be spent where they would get a bigger benefit for the dollar in the overall larger watershed. They used marginal economic analyses coupled with technical considerations of remedial options to demonstrate this practicality.

3.1.6 Finding S&T 6: UAA and TMDL Studies Were Most Effective When Integrated

Finding: UAA and TMDL studies were most effective when integrated. Successful UAAs that also involved a TMDL were usually conducted before the TMDL or in parallel with the TMDL.

Discussion: This, of course, is a logical finding given that TMDL targets are based on water quality standards, and thus a defensible TMDL begins with defensible standards. This is consistent with recent assessments of the TMDL program such as the National Academy of Sciences (2001), which recommends that states:

"…develop appropriate use designations for waterbodies in advance of assessment and refine these use designations prior to TMDL development. [Clean Water Act] goals of fishable and swimmable waters are too broad to be operational as statements of designated uses. Thus, there should be greater stratification of designated uses at the state level (such as primary and secondary contact recreation). The appropriate designated use may not be the use that would be realized in the water's predisturbance condition. Sufficient science and examples exist for all states to inject this level of detail into their water quality standards. To ensure that designated uses are appropriate, use attainability analysis should be considered for all waterbodies before a TMDL is developed."

This integration concept has been more recently advocated by U.S. EPA (King, 2006):

"**The UAA process should be better integrated with TMDL development**. We need to work together with states and tribes to ensure that as we develop TMDLs, we also coordinate on issues related to use attainability as needed. In practice, the information gathered to develop a TMDL, and the allocations in a TMDL, may point to the need to pursue a UAA. In some cases it may be more effective to ensure that the right uses are in place before completing a TMDL, however, it is also important not to let uncertainty about a specific water quality endpoint delay implementation of needed water quality improvements. Scarce resources should be directed where they will be most effective and avoid duplicative efforts. We should continue to share ideas/examples, develop and promote best practices."

Case Study Example: Louisiana and Maine UAAs

Louisiana was one of the first states to face third party lawsuits for not conducting TMDLs. Consequently, Louisiana Department of Environmental Quality and U.S. EPA moved forward with conducting TMDLs for 303(d)-listed waterbodies. Yet, in the instances where water quality standards were not appropriate or waterbodies were impaired because of natural conditions, LA DEQ first conducted UAAs to ensure that the TMDL would be based on a defensible and attainable target. If the waterbody was not attaining the use and the reason was one of the six U.S. EPA 40 *CFR* factors, then Louisiana Department of Environmental Quality conducted a UAA and refined the designated use before calculating a TMDL. Cases where a UAA was conducted in Louisiana include the New Iberia Southern Drainage Canal and Ramos Forested Wetlands at Amelia. Additionally, Louisiana refined the DO criteria to support the fish assemblage on an ecoregional basis.

Similarly, a UAA was conducted in Maine for both Ragged Lake and Seboomook Lake, two storage dams that prevent the upstream migration of invasive non-native species, maintaining the largest and least altered aquatic ecosystem in the Northeast. Although the waterbodies were listed as impaired, Maine believed it was not practical to conduct a TMDL until a UAA could be conducted that appropriately set the standards for the protection of upstream waters from non-native species.

3.2 Financial Needs and Impacts

In the category of findings related to financial needs and impacts, the findings can be grouped into two issues: (1) the economic justifications for UAAs; and (2) the cost of conducting the UAA studies. On the first issue, the review of case studies showed that economic affordability/feasibility and/or designated use valuations were seldom used. In part this was because guidance on these approaches was very limited and, when available, was overly narrow and controversial. On the second issue of cost to conduct UAAs, the case studies indicated a wide range of costs and that, at times, costs inhibited state involvement. The case studies also indicated that costs could be better contained when state protocols for UAAs are developed and followed. These findings are discussed individually below.

3.2.1 Finding F-1: Economic Affordability Was Seldom Used as Sole Justification

Finding: Economic affordability, that is substantial and widespread economic and social impact, was seldom used as the sole justification for the UAAs. Although cost was an important component of the decision criteria for many UAAs, an economic assessment following federal guidance was seldom presented as the UAA factor for change.

Discussion: The review of case studies found that most UAAs that have resulted in designated use refinements have not relied on 10(g) Factor 6 (socioeconomic impact). This is in part because the same six factors that are defined in U.S. EPA regulations for justifying a designated use change also are the factors for granting variances from water quality standards. As a result, it has been common for Factor 6 to be used by agencies to grant a variance, even if a designated use refinement may have been initially considered. This is often a matter of convenience or policy for the agency because a variance is explicitly intended to be temporary. Another reason for sparse use of Factor 6 is the recognition by UAA proponents that the federal guidance on this factor has substantial limitations and applicability, particularly for municipal entities, as described below for Finding F-2.

Case Study Exceptions:

Both the Spokane River UAA and Boston CSO UAA used the federal guidance to some extent, but also conducted additional economic analyses relevant to their situations. In the Boston Harbor CSO UAA, economic affordability was used (Factor 6) to successfully justify a use change allowing CSO discharges into Boston Harbor. However, upstream of the harbor the Charles River was given only a temporary variance because it was felt that additional information was needed to make a permanent use change there. The economic justification was not sufficient by itself. Further information on these case studies is provided in Chapter 5.0 and in the next finding.

3.2.2 Finding F-2: Federal Guidance for Economic Assessments was Rarely Used

Finding: Federal guidance for formal economic assessments was rarely used because it is narrow in interpretation and controversial.

Discussion: Federal guidance for Factor 6 (U.S. EPA, 1995) includes a two-tiered screening process to determine what level of pollution control would cause substantial impact (i.e., not be affordable) for a particular municipal entity. The first tier establishes that if controls cost less than 1% of median household income, then the controls are deemed affordable. If greater than 1% of median household income, then controls may or may not be affordable, depending on a variety of other considerations, although generally costs must exceed 2% to be potentially not affordable. Although 1– 2% may initially sound like reasonable thresholds, in practical reality they are difficult for many municipalities to meet this threshold because bond repayments can often be spread over a long period and a relatively large base of ratepayers. In addition, the use of such simple criteria applied to only median household income does not account for the economic impact to lower income households in a service area for which hundreds of dollars a year in increased sewer bills can have a substantial impact.

Case Study Example: Boston Harbor CSO UAA, Boston, Massachusetts

In this rare exception, Massachusetts conducted a formal Factor 6 UAA to justify a use change in areas of Boston Harbor affected by CSOs. Massachusetts used a detailed cost/benefit and affordability analysis, which included considerations that were not included in the federal guidance such as specific water and sewer issues, local social conditions, and cost versus benefit. However, to gain U.S. EPA approval, a separate analysis using federal guidance was also developed. Massachusetts was able to successfully show economic hardship using the federal guidance and complete the UAA based on a formal Factor 6 assessment for one subarea. Further information on this case study is provided in Chapter 5.0.

3.2.3 Finding F-3: Economic Valuations of Water Uses Were Not Seen in the Case Studies

Finding: Economic valuations of the designated uses were not observed as a method in any of the case studies reviewed to justify a change in waterbody use.

Discussion: Although economic valuation methods exist and are fairly well accepted for certain types of water uses (e.g., travel cost methods for recreation uses) they do not appear to have been used in any of the UAAs examined as part of this review. Such methods could be important in some future UAAs. For example, if attaining a certain aquatic life use, such as salmonid spawning, for a waterbody would require removal of a dam to provide suitable spawning habitat, then the relative economic value of fly fishing in a free-flowing riverine reach could be compared to the value of reservoir fishing.

Case Study Example:

No case studies were found in which an economic valuation of the use was conducted, therefore all of the case studies support this finding.

3.2.4 Finding F-4: UAAs Had a Wide Range of Costs, Methods, Complexities, and Schedules

Finding: Researchers observed a wide range of costs, complexities, and schedules associated with conducting a UAA, with the cost dependent on site-specific conditions. Large complex UAAs could cost hundreds of thousands of dollars and much more, while in some states streamlined simple UAAs conforming to standardized protocols could cost ten thousand dollars or less.

Discussion: The streamlined processes have often been developed out of necessity by various state agencies that have a mandate (e.g., legislation in Kansas) to complete numerous but fairly straightforward recreational UAAs in a relatively short period of time. Such streamlined protocols can facilitate recreational UAAs for mostly small streams in nonurbanized areas, and have generally been used only to differentiate between primary contact and secondary contact uses. On the opposite extreme are the very large, very complex UAAs like the Chesapeake Bay. However, most of the case studies investigated as part of this review fell somewhere in between these two extremes and are representative of a typical UAA for wet-weather, urbanized area, and effluent-dominated situations.

> Case Study Example:
>
> The range of costs is illustrated by the Santa Ana River UAA (California) and Chesapeake Bay UAA, which cost millions of dollars, versus low-cost, streamlined UAA protocols for recreation uses in Colorado, Kansas, and Missouri. In the case of Santa Ana River, the investment spent on the UAA (over $2 million in technical and legal fees) involved extensive data collection and modeling over a 5-year period and also extensive regulatory and legal negotiation. Still this was far below the anticipated treatment costs associated with compliance with the assigned uses ($2 billion in capital and operations and maintenance). In contrast, the Kansas recreational UAAs are conducted for just a few thousand dollars, if that, because they rely on a simple determination of existing uses derived from 1 or 2 days of field work and standardized but minimal supporting documentation. Further information on these case studies is provided in Chapter 5.0.

3.2.5 Finding F-5: Following State Guidance Minimized Costs and Streamlined the Process

Finding: Following state guidance when it was available minimized costs and streamlined the UAA process.

Discussion: This finding, of course, is applicable to agency-led UAAs in those states that have state-wide protocols (see Chapter 6.0). However, other stakeholders have also benefited from the availability of such state-wide and streamlined protocols.

> Case Study Example: Missouri Department of Natural Resources (MDNR) Recreation UAA Protocol
>
> MDNR proposed changes to the state's bacteria standards that would result in the use of whole body contact recreation (WBCR) being assigned to all classified waters of the state. This new use classification would require 911 discharging facilities to provide disinfection of their effluent at an estimated cost of $300 million statewide. The state applied its Recreational Use Attainability Analysis Protocol (MDNR, 2004) to 500 waterbodies (90% of the state's classified waters) where there were known facilities discharging bacteria to intermittent streams. All of the UAAs were conducted based on Factor 2 (low flow resulting in insufficient depth for WBCR). Municipalities and industries have conducted UAAs for the remaining 10% of classified waters. The UAAs were handled through the state's rulemaking process with the rules becoming final in September 2005. The protocol served the need for streamlining the UAA process and providing a well-defined approach for UAAs conducted by both state staff and other stakeholders. See Chapter 6.0 for more details on the Missouri protocol.

3.2.6 Finding F-6: State Budget Constrictions Inhibited State-led UAAs

Finding: State budget constrictions inhibited state led UAAs.

Discussion: Most state water quality agencies across the country are faced with shrinking or frozen budgets, coupled with increasing responsibilities such as completing numerous TMDLs in a short time frame in response to litigation. As a result, it was no surprise to learn in this research that most states, other than those mandated by legislation such as Kansas, have not directed substantial resources to doing UAAs or further refining their subcategories of use. As a result, a number of the recent case studies reviewed were UAAs that were sponsored by entities impacted by water quality regulations in situations in which it seemed intuitive that the water quality standards needed refinement to reflect existing and attainable conditions.

> Case Study Example: Old Alamo Creek, City of Vacaville, California
>
> The City of Vacaville, California, discharges into Old Alamo Creek, an effluent-dominated waterbody and tributary to a distant abandoned water supply intake. Downstream of the discharge, fish passage in Old Alamo Creek is barred by flapper gates. The City of Vacaville was having difficulty meeting, and could not economically justify meeting, the aquatic life uses and end-of-pipe drinking water standards so they moved forward with a UAA. To obtain the time, support, and attention of an under-budgeted regulatory staff, the city paid the regulatory agency for their time to work on the UAA.

3.3 Legal and Regulatory Requirements

In the context of legal and regulatory issues, the findings can be generally categorized into two categories: (1) follow the regulations; and (2) consider all stakeholder issues. The case studies indicated that carefully complying with all applicable regulations and thoroughly documenting that compliance through a complete administrative record was important to success and avoiding litigation. The case study review also indicated that it was important to consider all stakeholder issues and potential regulatory constraints by involving a coordinated stakeholder effort, considering downstream users, and anticipating the relationship to other regulatory programs like TMDL, CSO, and NPDES. These findings are discussed below.

3.3.1 Finding L-1: Compliance with State and Federal Rules and Procedures is Essential

Finding: The legal requirement in all UAAs was to comply with state and federal statutory and regulatory water quality requirements and follow applicable administrative procedures.

Discussion: As noted in Chapter 1.0, water quality standards are state and tribal rules, and thus any change to standards brought about by a UAA process must be formally promulgated by the state or tribe using the applicable administrative procedures. These adopted standards, however, also have to be approved by U.S. EPA. Therefore, it is essential to view compliance in context of both state, tribal and federal rules and procedures.

> Case Study Example:
>
> All UAA case studies for which revised standards were adopted by states have followed applicable administrative procedures.

3.3.2 Finding L-2: A Comprehensive Administrative Record Was Important

Finding: Building an administrative record was important and typically required extensive documentation of work plan development, agency meetings, and agency communications from the beginning of the process.

Discussion: The administrative record is the final backstop when disputes arise late in the UAA process. The review indicated that having solid documentation of agreements, analyses, compliance with procedures and rules, etc., was an important attribute in successful UAAs. This finding is most pertinent to nonagency sponsors of a UAA that may be required to file a formal petition for rule-making in the event that the agency itself does not support the UAA, or cannot conduct the UAA itself because of staff resource constraints.

> Case Study Example: Spokane River UAA
>
> In this case a consortium of dischargers conducted a UAA requesting changes to the designated uses of the Spokane River and Long Lake Reservoir and associated DO criteria. The need for the UAA was driven by an ongoing DO TMDL for the river and reservoir. The State of Washington did not want to adopt the recommendations in the UAA. However, extensive documentation and record-building, along with a thoughtful strategy, were included in this UAA process. As a result, the sponsors petitioned for rule-making based on their compliance with the law documented in the administrative record. Ultimately they prevailed when the state requested that the sponsors withdraw the UAA petition without prejudice in exchange for substantive negotiation of the DO TMDL. The subsequent agreement between the state and sponsors is that the UAA will be reconsidered at a later date as part of an adaptive TMDL implementation process. Further information on this case study is provided in Chapter 5.0.

3.3.3 Finding L-3: UAAs Are Usually Linked to Other Regulatory Programs

Finding: Unexpected legal obstacles can often occur because UAAs play a role in other regulatory programs (TMDL, NPDES, CSO, LTCPs, etc.).

Discussion: Potential litigation should be a real consideration, and UAA studies need to carefully follow regulations and thoughtfully integrate with other regulatory programs.

> Case Study Example: Multiple UAAs
>
> A large percentage of the UAAs that were examined as part of this research played a role in other regulatory programs. For example, of the 54 UAAs, 41% related to POTW discharges, and 17% related to CSO discharges.

3.3.4 Finding L-4: A Solid Technical Foundation Avoids Legal Challenges

Finding: A solid technical foundation is a key to avoiding legal issues and costs. Only one UAA has itself been litigated.

Discussion: It was surprising to the research team that among the case studies examined only one UAA was litigated. This may be due in part to the relatively small numbers of UAAs that have actually led to changes in use designations, except for those state-wide UAA programs described in Chapter 6.0 such as Kansas and Missouri. However, more litigation may occur in the future given the greater likelihood of more UAAs across the country as a result of increasing regulatory pressures in permits, TMDLs, and CSO programs. This finding may further explain why UAAs generally have not been litigated to date as discussed in Finding L-3. Most of the recent UAA processes have recognized the necessity for defensible technical data and analyses, as described above under Scientific and Technical Findings. It was observed that in all but the more simple case studies the technical foundation of data and analysis was well thought out and scientifically appropriate to justify the UAA.

Case Study Example- Cuyahoga and Ottawa River UAAs Contested

Although litigation has been rare, in the late 1980s two UAAs were contested in Ohio: the Cuyahoga River UAA (upstream of the ship channel); and the Ottawa River UAA. In both instances, the river was being upgraded from a limited warmwater habitat use designation, which has the least stringent requirements, to a warmwater habitat use, which requires significantly better water quality. Regulated entities contested the basis for the upgrades, challenging Ohio's use of the newly implemented numeric biocriteria and tiered aquatic life uses (more discussion in Chapter 6.0). The Ohio Supreme Court resolved the Cuyahoga River case, confirming Ohio's use of biocriteria and tiered uses for conducting UAAs. Subsequently, the Ottawa River UAA was settled, resulting in more stringent regulation of point and nonpoint sources.

No case study examples were explicitly identified where a strong technical foundation was clearly the reason that litigation was avoided. It is hard to establish a cause–effect relationship to an action/decision to litigate that did not happen, and explanations are not necessary. Nonetheless, the project team believes that in every instance where a strong technical case was built for the UAA, it contributed to a generally smooth process.

3.3.5 Finding L-5: UAA Coalitions were Common in Complex Urban Settings

Finding: In complex, urban-area UAAs, coalitions of stakeholders with solid legal support and well-coordinated legal strategies were more successful than UAAs with individual proponents.

Discussion: Several prominent UAAs supported this finding, including the Spokane River, Santa Ana River, Snake River, Chesapeake Bay, Colorado Recreational, Central Valley, Lower Boise River Tributaries, Ramos Forested Wetlands, and California Engineered Flood Channels UAAs.

Case Study Example: California Engineered Flood Channels UAA

State of California and regional regulators worked closely with stakeholders to revise recreational uses reflecting post-storm conditions. The State Water Resources Control Board directed the Los Angeles Regional Water Quality Control Board to review its recreational uses during wet weather when public entry is prohibited. Thus, the UAA process required a collaborative effort between staff from the Los Angeles Regional Water Quality Control Board (RWQCB) and other local agencies charged with protecting public safety including Los Angeles County's Multi-Agency Swift Water Rescue Committee. Although it is not possible to say if the process would have been successful without this collaboration, those involved felt that the collaboration was a key factor leading to success. Further information on this case study is provided in Chapter 5.0.

3.3.6 Finding L-6: Downstream Uses Were Not Always Fully Explored

Finding: 40 *CFR* 131.10(b) states that "the State shall take into consideration the water quality standards of downstream waters and shall ensure that its water quality standards provide for the attainment and maintenance of the water quality standards of downstream waters." However, the research found that downstream uses, such as intrastate, interstate, or tribal uses, were not always fully explored in the UAA process. When downstream uses were not considered, difficulties in finalizing the UAA or supporting a designated use change were sometimes encountered.

Discussion: Although the downstream uses issue was not prominent in most of the case studies that were examined, it was very important in several of them such as the Spokane and California Engineered Channels UAAs (see Chapter 5.0 for details). The research team believes that this will be an emerging UAA issue that will be increasingly important in future UAAs, especially the larger and more complex UAAs with a geographic scope that includes numerous downstream water users or tribes with their own water quality standards.

Case Study Example: Multiple UAAs

Many of the case studies demonstrated how downstream uses can be considered in a UAA. In the Fenhalloway (Florida) UAA, downstream waters were considered but were already designated as the use supported by the UAA for the upstream segment, so a use change would not have impacted downstream waters. For the New Iberia UAA in Louisiana, the use was changed upstream to secondary recreation, but dischargers still had to meet primary recreation standards to protect downstream uses. In East LA Lancaster, the stream was terminal so there were no downstream uses to explore. In other case studies, the stream discharged to the ocean (e.g., Boston Harbor, San Luis Obispo) and use changes were not thought to impact these waters.

In another example, Old Alamo Creek designated use was removed for municipal and domestic water supply, cold freshwater habitat, migration, and spawning. The use changes were supported by a UAA led by the City of Vacaville, California, whose wastewater is discharged into Old Alamo Creek. When adopting the designated use change, the State of California did not provide direction regarding the designated uses of receiving waters downstream of Old Alamo Creek. Unfortunately, the City of Vacaville's discharge violates the municipal and domestic water supply designated use of downstream waters, and now options (e.g., a UAA for downstream waters, site specific objectives, etc.) are being evaluated for the downstream waters.

3.4 Regulatory Agency Preparedness

The preparedness of regulatory agencies to openly consider UAA processes and their abilities to understand the requirements was deemed critically important to the success of UAAs. Findings in this research showed that agencies were most prepared if they had already developed guidance or protocols and were least prepared when faced with complex situations. Difficulties also arose when the state agency did not fully anticipate the involvement of other agencies like U.S. Fish and Wildlife Service or U.S. EPA Regions. In addition, agencies were often not initially open to consideration of alternative solutions, but when the agencies were effectively engaged these alternatives provided satisfactory outcomes for all parties.

3.4.1 Finding R-1: State Protocols and Guidance Support Regulatory Preparedness and a Smooth Process

Finding: When statewide UAA protocols and/or guidance exist, state agencies tended to be well-prepared and the UAA process proceeded more smoothly.

Discussion: This finding is largely predicated on the success of statewide program guidance and methodologies in Kansas, Oklahoma, Missouri, Ohio, Minnesota, and Colorado. In these states more UAAs were approved at lower cost and with fewer disputes (see Chapter 6.0). Few other states have UAA guidance and protocols but a number are in the process of developing guidance (e.g., Oregon and Washington) or are considering doing so (e.g., Idaho).

> Case Study Example: Oklahoma Statewide UAA Protocol
>
> Oklahoma has developed the "Unified Protocols for Beneficial Use Assignment for Oklahoma Wadeable Streams," which provides procedures for streamlined field assessments and follow-through procedures. The result is more than 30 UAAs completed in the last 10 years and over 100 in the last 20 years, all without litigation or administrative appeal. The majority of the UAAs were conducted quickly and relatively inexpensively. Overall, the efficiency in the established protocols and guidance allowed for not only money to be saved in conducting the UAA, but also money to be saved by dischargers when planning for the correct level of protection early on.

3.4.2 Finding R-2: States Are Less Prepared in Complex UAAs

Finding: In more complex UAAs or in states with limited experience, state agencies are less prepared unless proponents initiate close cooperation.

Discussion: This finding closely parallels Findings F-6 and L-5. States are intrinsically cautious about UAAs, but in the face of a complex technical and political UAA, state agencies were generally much more cautious if not reluctant.

> Case Study Example: Snake River-Hells Canyon UAA
>
> State agencies from Idaho and Washington and U.S. EPA were focused intently on a complex TMDL (eight pollutants for 200+ river miles of the Snake River, the largest river in Idaho). Initial discussions with the various agencies revealed reluctance to embark on a UAA given limited State resources and an intensive TMDL schedule. As a result, a large coalition of stakeholders initiated the use refinement process in parallel with the TMDL hoping that the results would be incorporated into water quality objectives of the TMDL. At the time, only a handful of UAAs had been conducted and approved in Region 10, with no use analyses on the highly modified and complex Snake River system. A series of meetings was coordinated with the State and stakeholders to discuss UAA methods and potential outcomes. As a result of the coalition and close coordination with the two states, the TMDL explicitly recognized the UAA effort and noted that refinement of designated uses may be appropriate for some segments of the river and should be further considered as part of the adaptive implementation process established by the TMDL.

3.4.3 Finding R-3: Fish and Wildlife Agencies Are Important to UAA Approval

Finding: The input of federal and state fish and wildlife agencies is important to Aquatic Life UAA success.

Discussion: On the state level, fish and wildlife agencies often are in different departments or at least different divisions of government than the water quality agencies. The goals and programs of each agency may not be consistent or in total alignment in all cases and it is important in the UAA process to understand this. For example, the water quality regulations may be intended to protect only native, indigenous fish species, whereas the fisheries agency may have management goals that include non-native or introduced game fish as a key component of their program for certain waterbodies. For example, in the Pacific Northwest only coldwater fish species are native, but cool- and warmwater fish have become important game fish with explicit management goals in major reservoirs and lakes. In addition, the federal agencies have important responsibilities in all states, but in particular in situations in which federally threatened and endangered species are involved.

> Case Study Example: Harrisonburg (Virginia) Wastewater Treatment Plant Chlorine Variance UAA
>
> The Virginia Department of Environmental Protection (VDEP) conducted a UAA on a small stream receiving wastewater treatment plant effluent from a small community. Without costly additional treatment works, the effluent would not be able to meet the existing instream standards and criteria for chlorine. The VDEP UAA concluded that very limited fish populations existed in the streams because of low-flow conditions, and that aquatic life was not an existing use. The U.S. Fish and Wildlife Service weighed in heavily on the determination of "existing" use, arguing that any fish present in pools constitutes an existing use, versus the VDEP approach that a "balanced" community must be present. U.S. EPA subsequently did not approve the UAA. Had this issue been considered early with the U.S. Fish and Wildlife Service in the investigative process, suitable analysis might have been conducted or, alternatively, a different regulatory strategy might have been pursued to the satisfaction of all parties.

3.4.4 Finding R-4: U.S. EPA Regional Approval Is Required but Experience with and Support for UAAs Varied by Region

Finding: U.S. EPA regions are the ultimate decision-makers for approval of designated use changes based on UAAs. The research found that experience with UAAs and support by U.S. EPA for changing uses varied widely by U.S. EPA region and the individuals involved.

Discussion: As described in Chapter 1.0, any changes to state water quality standards must be formally approved by U.S. EPA before they can be used for Clean Water Act purposes. Compilation of the selected case studies for this research revealed that numerous UAAs have been conducted and approved in U.S. EPA regions 5, 6, 7, and 8 in large part because of state-wide processes in Ohio, Oklahoma, Missouri, Kansas, and Colorado. Some of the other U.S. EPA regions have fewer approved UAAs and/or require more rigorous demonstration. This is partially caused by the absence of standardized protocols or guidance, making their review more cautious, time consuming, and at times requiring a higher degree of justification.

Case Study Example: Multiple UAAs

From the review of more than 200 case studies, the research team has found that experience in conducting a UAA is not the same from one U.S. EPA region to another, nor is the effort involved, or the support by the U.S. EPA regional office. Although UAAs are a means to appropriately designate uses, the actual data needed and process to conduct a UAA has not been systematically standardized by U.S. EPA guidance. Consequently, each U.S. EPA region has used best professional judgment as to how to interpret the data needs and the hurdles to cross to support a use change. It was observed that the level of effort, length of time, required data, in addition to the attention given to UAAs can vary widely. For example, Oklahoma and Region 6 generally have a cooperative working relationship in reviewing standards changes because Oklahoma has a standardized protocol. Also, Region 5 has had a cooperative working relationship with Ohio in reviewing standards changes supported by UAAs because of a highly developed bioassessment program. However, the absence of standardized protocols and federal guidance in this area has made some regional reviews much more cautious. U.S. EPA headquarters has taken recent action to encourage consistency and a more open dialogue regarding UAAs among the regions.

3.4.5 Finding R-5: Alternative Regulatory Solutions Were Often Advantageous

Finding: Some UAAs were resolved by alternative regulatory solutions, not UAA approval. New designated use categories and temporary standard applications have resulted from the UAA process.

Discussion: The most common alternative regulatory solutions that evolve from explicit UAA processes include variances (see discussion for Finding F-1) and site specific criteria. However, there are a number of other regulatory alternatives that are commonly employed in Clean Water Act processes such as mixing zone studies and water quality credit trading. The reader is encouraged to read the UAA Handbook cited in Chapter 1.0 (WERF & NACWA, 2005), which has a separate, detailed appendix that describes the elements of each of these alternatives, some challenges and advantages of each, and specific real world examples.

Case Study Example: Chesapeake Bay UAA

This extensive, multi-state, multi-U.S. EPA Region UAA established several new subcategories of aquatic life use designations for the Bay, including the specific season in which each of these uses occurs. At the start of the process more restrictive NPDES permits were being proposed for nitrogen discharges from area wastewater treatment plants. Extensive data collection and modeling demonstrated that the existing standards were unattainable. As a result of negotiation and more technical analysis, a UAA was undertaken which resulted in stakeholder approval of entirely new designated uses for the Bay with phased implementation. In addition, on the regulatory side, technology requirements were imposed at municipal treatment plants but without initial effluent limitations. Also, an adaptive management strategy was implemented to re-evaluate and revise the plans over time. These innovations were not anticipated at the start of the process but came about through collaborative scientific examination of the situation. Further information on this case study is provided in Chapter 5.0.

3.5 Public Awareness and Involvement

Public awareness and involvement are elements of high importance in UAAs, not only because of regulatory requirements, but more so because a change in use represents a perceived change in a public resource and regulatory agencies place high value on public reaction. Based on the review of case studies, the research team found that public involvement varied widely in successful UAAs but when successful it generally involved more than the minimum required effort and included face-to-face communication. Two important points were found to be critical to the public education message: (1) the UAA process was inherently an objective process to assess the actual attainable uses, not an effort to degrade water quality and/or downgrade uses; and (2) the choice of uses can have significant additional economic costs. Finally, it was observed that politicians generally do not get involved in UAAs except when large expenditures are involved and then, typically, they were supportive of the process. Below is a summary of those findings.

3.5.1 Finding PA-1: Public Involvement Typically Exceeded Regulatory Requirements

Finding: Successful UAA case studies always followed the minimum requirements for state rulemaking guidelines but, in general, exceeded those requirements with considerably more public involvement.

Discussion: Each state has its own rule making requirements for public involvement (e.g., California Environmental Quality Act, Massachusetts Environmental Policy Act, etc.) These administrative procedures require a minimum of public notice and formal comment periods. However, most successful efforts included considerably more public involvement than the minimum. The review of case studies found that the minimum effort was not sufficient, especially for the more complex or controversial UAAs.

Case Study Example: Multiple UAAs
All UAA case studies for which revised standards were adopted by states have followed applicable administrative procedures including provision of public notice comment opportunities.

3.5.2 Finding PA-2: Public Involvement Varied Depending on Site Complexity

Finding: The level of public involvement typically varies based on the complexity of the UAA. UAA complexity is often dependent on factors such as crossing state or tribal boundaries, how urbanized the area is, geographic scope, and the value or use of the resource.

Discussion: This finding is fairly self-evident and does not require substantial discussion. The obvious range in scope extends from the extensive Chesapeake Bay UAA to the more limited streamlined UAAs in states such as Kansas.

Case Study Example: Kansas City Recreational UAAs and Santa Ana River UAA

In situations like the Kansas recreational UAAs, the public involvement process was minimal because the analyses were relatively simple and the State elected to publish all of the UAA outcomes for a given year into one annual report. As with all rulemaking in the state, the public had a chance to comment on the annual report and provide input to technical staff. On the other side of the spectrum, for the Santa Ana (California) UAA, because the UAA affected nearly one million rate payers, the UAA proponents spent over $100,000 on public awareness activities that included multiple public workshops, development of an educational video, and a facilitated stakeholder group that was open to the public.

3.5.3 Finding PA-3: UAAs Were Often Misperceived as an Effort to Degrade Water Quality

Finding: Without education at the beginning, UAAs were often "misperceived" or mischaracterized as being a process to allow degradation of current water quality conditions, rather than an objective assessment of attainable uses beyond current water quality and existing uses.

Discussion: Clean Water Act programs such as TMDLs, NPDES wastewater and stormwater permits, and CSO LTCPs all are designed to protect and often improve water quality conditions relative to current conditions. The only question is how much improvement is needed to attain desired uses, and when are expenditures a misallocation of public resources if the additional requirements do not result in greater aquatic life or recreational use. As described in Chapter 1.0, U.S. EPA regulations stipulate that a designated use that is an existing use simply cannot be removed. Thus, UAAs are not intended to degrade water quality conditions, but simply help us understand and define what is feasible (attainable). Nonetheless, the review of case studies for this WERF research has found that misperceptions and mischaracterizations are fairly common, particularly by some environmental advocacy organizations. Guidance on developing effective stakeholder involvement processes is provided in Chapter 3.0 of the UAA Handbook (WERF & NACWA, 2005).

Case Study Example:

No specific case studies reviewed by the research team can be cited as examples, as this finding can often stop a UAA from moving beyond the proposal phase. There are constituents and stakeholders with nationwide presence that view UAAs overall as "… making it acceptable to add more pollution (Kruzen, 2005)." This stresses the importance of public involvement and education to ensure a proper understanding of the goal and benefits of each UAA.

3.5.4 Finding PA-4: Public Involvement Was More Effective When Face to Face

Finding: Face-to-face communication with the public is more effective than other mechanisms of communication.

Discussion: Although the research has found that other forms of public involvement are common and somewhat effective (e.g., websites, fact sheets, mailings, editorials, etc.), public involvement was most effective when the public was directly engaged face to face with the agency personnel and scientists involved. Successful UAAs invariably made significant efforts to organize educational meetings and workshops with diverse organizations (e.g., watershed groups, chambers of commerce, lake associations, etc.) to present and discuss the reasons for and methods of the UAA. This generally did not require a substantial level of effort and was typically included in the more complex and successful UAAs.

> Case Study Example: California Engineered Flood Channels UAA
>
> Because this UAA was sponsored, in effect, by the State of California as a modification to the Los Angeles Regional Water Quality Control Plan, the minimum public involvement requirements were followed. Above the minimum requirements, given the extensive agency coordination that was required, this UAA also included a public workshop and multiple open board meetings. Further information on this case study is provided in Chapter 5.0.

3.5.5 Finding PA-5: Involvement by Politicians Was Rare and Generally Supportive

Finding: Generally, there was not a high level of involvement by politicians either in favor or in opposition to noncomplex UAAs. When politicians were involved with UAAs, they generally supported the UAA.

Discussion: Successful, complex UAAs in urbanized areas have benefited from political support at the local (and state) government levels. The research has found that when these local officials were informed and educated about the UAA process (i.e., that the goals are to improve water quality and achieve feasible/attainable uses of the waterbody; not to degrade existing quality and uses), then they supported the process.

> Case Study Example: Indiana CSO Legislation
>
> The State of Indiana has 107 permits addressing CSOs. The CSO communities have been working cooperatively with the Indiana Department of Environmental Management (IDEM) in trying to create revisions for water quality standards that are appropriate for CSO impacted waters through legislative changes in the standard. These exceptions are necessary because Indiana's water quality criteria require that *Esherichia coli* concentrations not exceed 235 counts per 100 mL at *any time,* making it difficult to prepare a CSO LTCP that would comply with water quality standards at all times and in all places. With the support of State representatives and senators, the Indiana General Assembly enacted Senate Enrolled Act (SEA) 431 in 2000 which provided a format for coordinating CSO LTCPs with any necessary water quality standards reviews, and established a process for temporarily suspending the full body contact recreation after the LTCP is implemented. By continuing to work with the politicians, deficiencies in SEA 431 were identified and clarified by interested parties in 2005. In that year, SEA 620 was passed, creating a CSO wet-weather limited use subcategory in Indiana's recreational standards. This subcategory was more acceptable to IDEM and all stakeholders than the temporary suspension of standards.

3.5.6 Finding PA-6: Pressing Schedules and Large Potential Expenditures Were Motivators for UAA Analysis

Finding: The risk of spending large amounts of money in the face of tight TMDL schedules and NPDES requirements created an urgency or willingness to move forward with UAA analyses.

Discussion: The research shows that the risk of large expenditures in the face of urgent TMDL decision-making has been an important motivator for most of the nonagency sponsored UAAs. However, this same risk pertains to state water quality agencies as well and has led to several of the statewide streamlined approaches described in Chapter 6.0.

Case Study Example: Kansas Recreational UAAs
Kansas was facing a tight schedule for completing recreational TMDLs, driven by court-imposed deadlines. If the State of Kansas had elected to quickly conduct and approve inadequate TMDLs with unattainable standards, the result could have been onerous and expensive requirements for municipalities with little or no benefit, as well as potential costly and extended TMDL lawsuits. Thus, the state developed streamlined protocols for conducting hundreds of UAAs. Further information on this case study is provided in Chapter 5.0.

CHAPTER 4.0

GUIDANCE ON CHARACTERIZING AND CONDUCTING SUCCESSFUL UAAS

4.1 Introduction

The purpose of this chapter is to provide guidance on characterizing and conducting successful UAAs by translating the research findings from Chapter 3.0 into lessons learned. The goal of this chapter is to illustrate the deliberation process that is most likely to lead to an informed understanding of the UAA, and not attempt to ensure that each UAA will result in a change of use designation. Therefore, a successful outcome in the context of this guidance also includes situations where alternative regulatory solutions result, or situations where the UAA process leads to widespread stakeholder, public and political recognition that uses are either appropriate or need change. The successful UAA process is where the complexities are both recognized and understood and are directly considered in the decision.

In Chapter 3.0, factors for success focused on five broad categories of criteria. In this chapter, these criteria for success are incorporated into integrated guidance. In general, this integration means that there needs to be a shared vision for the use of the waterbody—even if the interests of every group cannot always be fully accommodated. A shared vision means a common perception of broad, long-range goals. Also, explicit coupling of the five integrated criteria is needed so no one group of viewpoints or issues dominates another and stakeholders feel they have input to the decision making. This integrated approach is a critical part of a successful UAA because each of these criteria works together and supports the UAA process, as shown in Figure 4-1.

Figure 4-1. Integrated Framework for a Successful UAA.

This chapter first provides a summary of the findings and recommendations from each of the five categories of criteria in Figure 4-1. A discussion of how these recommendations are integrated into a successful process is then presented. The chapter concludes with hypothetical examples illustrating the process.

4.2 Translating Findings into Recommendations

A discussion of the 29 findings that were derived from the detailed evaluation of the case studies is provided in Chapter 3.0. Observations were extracted by examining the answers to the research questions in the five categories of criteria:

- Scientific and Technical Understanding (S&T);
- Legal and Regulatory Requirements (L);
- Financial Needs and Impacts (F);
- Public Awareness and Involvement (PA); and
- Regulatory Agency Preparedness (R).

These observations or findings in themselves provided significant insights as to the factors contributing to success. However, to be more clear and explicit, specific recommendations were developed in response to each finding that the researchers felt were practices that would promote success. Tables 4-1 through 4-5 use a graphical format linking findings to recommendations, one table for each categorical criterion. A discussion is provided in subsequent sections of this chapter relating how these recommendations would be integrated into an overall process. The success of the UAA process may rest not just on the technical analysis, but also on whether it has successfully and broadly shared adequate information to generate interest, credibility, and consensus. The science of the watershed biology and chemistry needs to be aligned with the science of human behavior to gain success. Ideas about how to address this alignment are discussed in the sections below, beginning first with a discussion of the identification of stakeholders, followed by a discussion of recommendations derived from this research.

Table 4-1. Crosswalk Between Findings (Criteria for Success; Chapter 3.0) and Recommendations (Guidance; Chapter 4.0), Scientific and Technical Understanding.

Scientific and Technical Understanding (S&T)

	Finding	Recommendation
S&T 1	Recent successes typically had agency involvement throughout the process	*Develop a mutually agreed upon strategy with relevant agencies and stakeholders from the start.*
S&T 2	For stakeholder sponsored UAAs, agencies often requested more data than anticipated.	*Identify applicable factors, data needs, and critical terms with all agencies from the start.*
S&T 3	Successful UAAs generally had a decision pathway upfront relating critical definitions with applicable factors and data to specific outcomes.	*Define a decision pathway and data needs.*
S&T 4	Models are usually needed in complex systems to answer the attainability question for aquatic life uses.	*Use models to help answer attainability questions in complex systems.*
S&T 5	Consider cost-benefit/ effectiveness analyses to help define "feasible" for technical factor UAAs.	*Consider including economic analysis (e.g. cost-benefit or cost-effectiveness) to help assess "feasibility" for technical factor UAAs.*
S&T 6	UAAs and TMDL studies were most effective when integrated.	*Integrate with TMDL process from the start.*

Table 4-2. Crosswalk Between Findings (Criteria for Success; Chapter 3.0) and Recommendations (Guidance; Chapter 4.0), Legal and Regulatory Requirements.

Legal and Regulatory Requirements (L)

	Finding	Recommendation
L 1	The legal requirement in all UAAs was to comply with state and federal statutory and regulatory water quality requirements and follow applicable administrative procedures.	*Know the rules, identify gaps, appreciate schedules, and comply.*
L 2	Building an administrative record was important and typically required extensive documentation of work plan development, agency meetings, and agency communications from the beginning of the process.	*Keep good documentation (document agreements on critical terms such as "natural background," "feasible," "non-remedial," and irreversible"). Carefully coordinate scientific and technical reports and memoranda.*
L 3	Because UAAs play a role in other regulatory programs (TMDL, NPDES, CSO, LTCP, etc.) unexpected obstacles can occur	*Identify potential challenges and future issues related to other regulatory programs, as applicable, and develop strategies to address these.*
L 4	A solid technical foundation avoids legal issues and costs. Only one UAA has itself been litigated.	*Use sound science and a thoughtful process to avoid litigation.*
L 5	In complex, urban-area UAAs, coalitions of stakeholders with solid legal support and well-coordinated legal strategies were more successful than UAAs with individual proponents.	*Identify partners and develop coalitions early.*
L 6	Downstream uses (intrastate, interstate, or tribal) were not always fully explored.	*Identify and involve downstream users.*

Table 4-3. Crosswalk Between Findings (Criteria for Success; Chapter 3.0) and Recommendations (Guidance; Chapter 4.0), Financial Needs and Impacts.

Financial Needs and Impacts (F)

	Finding	Recommendation
F 1	Economic affordability was seldom used as the sole justification.	*Include cost/benefits considerations in factors other than Factor 6.*
F 2	Federal guidance for formal economic assessments was rarely used because it was narrow in interpretation, and controversial.	*Additional economic guidance for UAAs is needed.*
F 3	Economic valuations of the designated uses were not present in the case studies reviewed.	*Consider whether putting an economic value on the use is needed.*
F 4	UAAs had a wide range of costs, methods, complexities, and schedules.	*Scope the data and modeling needs upfront to know what the UAA will cost. This includes working with the state to define data requirements.*
F 5	Following state guidance generally minimized cost and streamlines the process.	*If guidance in your state is lacking, it may be beneficial to form coalitions to promote the use and/or development of state guidance even if from other states.*
F 6	State budget considerations inhibit state led UAAs.	*Consider neutral parties to lead the facilitation of UAAs.*

Table 4-4. Crosswalk Between Findings (Criteria for Success; Chapter 3.0) and Recommendations (Guidance; Chapter 4.0), Public Awareness and Involvement.

Public Awareness and Involvment (PA)

	Finding	Recommendation
PA 1	Successful UAAs typically included extensive public involvement beyond just regulatory notice.	*Do more public involvement than the minimum required by statute to obtain input from stakeholders.*
PA 2	The level of public involvement typically varies and is dependent on the complexity.	*Identify issues early and develop a stragegy.*
PA 3	UAAs were often "misperceived" as an application to allow degradation of current water quality.	*Educate the stakeholders on purpose and potential outcomes of UAA.*
PA 4	Public involvement was more effective with face-to-face communication.	*Use face-to-face communication to inform stakeholders of the UAA and its results.*
PA 5	Involvement by politicians was rare but generally supportive.	*Communicate issues to decision makers. Define economic impacts for elected officials.*
PA 6	Pressing schedules and large potential expenditures were motivators to undertake UAA analysis.	*Identify the economic investments at stake early and educate the public of the consequence.*

Table 4-5. Crosswalk Between Findings (Criteria for Success; Chapter 3.0) and Recommendations (Guidance; Chapter 4.0), Regulatory Agency Preparedness.

Regulatory Agency Preparedness (R)

	Finding	Recommendation
R 1	Using state UAA protocols and/or guidance supported regulatory preparedness and a smoother UAA process.	*Make sure you understand state rules/procedures, UAA guidance, and rulemaking schedules.*
R 2	States are less prepared in more complex UAAs or in states with limited experience.	*Work with agencies to define the scope by making use of any available guidance.*
R 3	Federal and state fish and wildlife agencies are important to aquatic life UAA success.	*Work with fish and wildlife agencies to understand/address their concerns.*
R 4	EPA regional staff approval is required but support for UAAs varies by EPA region.	*Identify EPA Regional issues early and work into your strategic plan.*
R 5	Alternative regulatory solutions such as new designated use categories and temporary standards applications were often advantageous.	*Be open and evaluate alternative regulatory outcomes.*

4.3 Identifying the Stakeholders

A successful UAA will depend on the support of multiple stakeholders and, therefore, a successful effort must start with broad stakeholder considerations. Everyone who has an interest in the watershed should be invited to participate in the development and implementation of the UAA. One cannot be too inclusive. Parties with an interest can always decline to participate or limit their involvement. But failure to invite and engage a critical stakeholder early in the process can doom the UAA to failure.

It should be understood that uneducated or misinformed critics can undermine and sidetrack the group process and that different groups may have already framed decisions regarding the UAA. Different strategies may be needed to engage different types of stakeholders. Many stakeholders can be engaged through broad public meetings, newsletters or news articles. However, key stakeholders capable of significantly impacting the UAA process, whether constructively or detrimentally, should be addressed using a stakeholder specific strategy. For example, one strategy would be for program sponsors to meet with selected individuals or organizations in a smaller setting for the purpose of conducting outreach and education, before inviting them to join a larger group or a steering committee. In this setting information exchange can be tailored to meet individual needs in an informal manner, rather than in a public debate. In addition, these smaller settings facilitate relationship building which cannot be accomplished in larger public meetings or late in the process. Good relations lead to a spirit of cooperation rather than conflict and mistrust.

The success of the UAA may often rest on whether it follows an inclusive outreach program designed to produce broad information and generate a collaborative approach -- an approach that does not have a predisposed point of view at its inception. This is essential to gain the trust of stakeholders. Stakeholders, including agencies, need to feel that the UAA sponsors sincerely want their input to inform the outcome, and that this outcome is not already predetermined. The stakeholder education process must focus on full and comprehensive disclosure, active engagement, and openness to different points of view.

Involving Stakeholders: ***Key questions in designing your process***. The agency promoting the UAA should consider these questions before developing the plan to involve stakeholders:

Are there individuals, organizations, and key stakeholder groups who are anticipated to oppose the UAA project? What about those who would support it? Invite <u>both</u> groups.

Do you know the priorities and key concerns of the community you serve and the stakeholders who will participate? Directly address these issues in your process.

Are there unmet needs and/or real or perceived problems that should be addressed? Are these problems commonly understood?

Is there a level of basic trust among the required participants? Will your outreach plan benefit from the help of a professional facilitator, mediator or conflict resolution expert?

Is there understanding of the needs in the watershed? If the UAA is viewed as a way to satisfy commonly held needs, then this knowledge can generate a motivation for moving forward.

How welcoming is the regulatory environment to the development of the UAA? How should their participants be included in the stakeholder outreach plan?

Do stakeholder expectations for the watershed reflect actual land use, geology, topography, and hydrology or unrealistic expectations?

Is there a shared vision for the watershed? Is the potential UAA outcome compatible with this stakeholder vision? How was the shared vision developed and was it a product of collaboration, coupled with strong supporting science?

If you need to develop a shared vision have you thought through the process from beginning to end including how you will educate and engage all stakeholders, how you will mediate conflicting views, and how you will incorporate good science?

4.4 Integrating Recommendations into the UAA Process

The tendency to think about and focus on each individual recommendation in this report independently is enticing because of the diversity in the UAA process and the potential complexity of each specific area. However, to be successful, it is necessary to integrate all of the recommendations in each phase of the UAA process and fully consider stakeholder issues at each step. Technical information and results of scientific analyses are important, but process and effective stakeholder involvement are also keys. If one approaches the UAA with a foregone conclusion, however strong and scientifically based, there is a tendency to polarize debate, question motivations and drag the process into paralysis. It is important that the process be open and thoughtful and the sponsors carefully consider all recommendations from this report. In UAAs, process is as important as substance. For purposes of discussing recommendations in an integrated plan, the UAA process has been divided into four phases:

- Getting started;
- Conducting the UAA;
- Involving the public and interest groups; and
- Resolution.

Table 4-6 summarizes how the researchers see the integration of the recommendations into the UAA process involving these four phases. The criteria and recommendations for each phase are discussed in the sections following the summary table.

4.4.1 Getting Started

The majority of recommendations are best implemented at the beginning of the UAA process, primarily because starting the UAA process "right" tends to lead to more successful UAAs. Investments at the beginning of the effort ultimately leads to a more efficient and effective UAA. "Right" in this context means investing resources and time into four main areas:

- Determine who should be involved, who should lead the UAA, and keep decision-makers informed and engaged;
- Develop a mutually agreed-upon strategy to satisfactorily address potential issues for *all* relevant agencies and stakeholders; and
- Define a decision pathway and the required data needs.

Each of these steps is described below.

4.4.1.1 Involve All Affected Stakeholders

UAAs require close cooperation between regulators, those who are regulated, and all affected stakeholders independent of who initiates the UAA. This should involve regulators, permit holders, water users, relevant NGOs, and the general public. UAAs that are undertaken in cooperation with a variety of interests and coalitions tend to be more broadly supported, leading to more successful outcomes. Demonstrated capacity to work collaboratively toward a common vision enhances prospects for success. No deliberation and public involvement approaches can be successful until all potential conflicts among all stakeholders are recognized and addressed.

Because UAAs may be complex, costly, and sometimes controversial, it is important to understand the stakeholder environment, including cooperation with regulatory agencies, elected officials, and nongovernmental organizations. Economic considerations from a cost-benefit perspective also need to be understood so that an informed decision to proceed is possible. Understanding the issues and relationships early can be critical to developing a successful process and strategy. For example, mediation of prior differences might be the most appropriate strategy before initiating the UAA stakeholder process. Also, if stakeholders understand but are not ready to undertake and support a UAA process other alternatives should be evaluated.

Table 4-6. UAA Integrated Recommendations.

Getting Started
1. Identify and involve all affected stakeholders
a. Establish decision process ground rules for the role and expectations of the stakeholder process (S&T-3, PA-1)
b. Discuss whether there is a need for impartial third party facilitation or mediation (PA-2, PA-4)
c. Identify partners and collaborate on the identification of common issues – this can help to reduce costs and leverage scarce resources and build support (L-5, PA-5)
d. Use a collaborative process to promote the use and/or development of state guidance even if from other states (F-5, R-1)
e. Involve all regulators early in the process, learn their issues and make certain that there is a willingness to comply with rulemaking schedules and guidance (S&T-1, L-1; R-1)
2. Develop a work plan and strategy for the UAA
a. Work with all relevant agencies and stakeholders to understand all the issues and regulatory constraints (PA-2, PA-3, L-3)
b. Develop a mutually agreed upon strategy that addresses all issues and identified constraints (S&T-1, PA-2, R-3, R-4)
c. Integrate the UAA with other regulatory programs and agencies (S&T-6, L-3, R-3)
d. Engage U.S. EPA Regional staff early on (R-4)
3. Develop a decision pathway
a. Define the decision pathway and relevant data needs (S&T-3)
b. Identify and define applicable factors and critical regulatory terms (S&T-2, L-1, L-2)
c. Work with agencies to define the scope, by making use of any available guidance (R-2, S&T-2)
d. Scope the data and modeling needs upfront to know what the UAA will cost (F-4, S&T-2)
e. Use models to help answer attainability questions in complex systems (S&T-4)
f. Identify and involve downstream users (L-6, PA-2)
Conducting the UAA
1. Work to gain approval
a. Use sound science and couple it with a thoughtful, inclusive stakeholder process (S&T-1, S&T-2, L-4, PA-1)
b. Keep good documentation (L-2)
c. Request mediation if litigation seems a likely result of conflict (PA-1, PA-3, PA-6)
2. Develop thorough and appropriate economic analyses (as needed)
a. Consider including economic analysis (e.g. cost-benefit or cost-effectiveness) to help stakeholders understand what may be "feasible" for technical factor UAAs (S&T-5, PA-5, PA-6)
b. Conduct an "affordability" evaluation if Factor 6 is a legal basis for the UAA; consider other indicators of affordability beyond U.S. EPA economic guidance (F-1)
c. Additional economic guidance for UAAs is needed (F-2)
d. Consider whether putting an economic value on the use is needed (F-3)
Involving Stakeholders
1. Fully engage the public
a. Do more public involvement than the minimum (PA-1)
b. Use face-to-face communication (PA-4)
c. Inform stakeholders on the purpose and potential outcome of UAA (PA-3)
d. Identify the economic investments at stake and educate the public of the consequences (PA-6)
2. Keep your decision-makers in the loop
a. Communicate the issues to real decision-makers (R-3, R-4, PA-5)
b. Define economic impacts for elected officials (PA-6, S&T-5)
Resolution
1. Be open to alternative regulatory outcomes (R-5)

Whether a coalition or a single entity elects to begin the UAA process, empowering a neutral party or mediator can be useful to facilitate the UAA process, help with discussions, and promote consensus agreements among parties (F-6). Bringing in a neutral party may help establish credibility with both the regulators and those who are regulated, including the public. Without this type of broad or neutral involvement, the UAA process is commonly viewed with skepticism and can often be derailed because of distrust that is characterized by frequent requests for more data and analysis.

Regardless of who leads, conducts, or facilitates the UAA process, it is essential to have all relevant stakeholders informed if not involved in the process. A UAA involves many diverse stakeholders including regulators, permit holders, water users and the general public. Understanding all of their interests and engaging all these stakeholders in the process is a significant factor leading to success. In this way the scope of investigations and process for decision making can be constructed at the beginning with elements that address critical issues for all stakeholder groups. The unfortunate alternative is to neglect some views and then later find the UAA process totally derailed because data and analysis were unavailable to address legitimate concerns. It is better to know the issues at the beginning and design appropriate studies than gamble on an unobstructed process.

Regulatory staff also should coordinate closely with their legal departments and affected stakeholders at the beginning so that potential implications of any outcomes of the UAA are well understood. This could range from changes in NPDES requirements to changes in anticipated uses. In addition, state regulators should coordinate closely with federal regulators to avoid differing interpretations and applications of the Clean Water Act and to consider the applicability of other federal programs. This vertical and horizontal coordination of agencies is important to ensure that there is a complete knowledge base of all applicable state water quality standard rules (L-1, R-1). Furthermore, understanding and following any available UAA guidance, state or regional, can be very beneficial to gaining legal standing as well as stakeholder support (R-1).

Regardless who leads the UAA process, it can often be useful to form coalitions and partnerships based on common issues. This can be beneficial in reducing overall costs and leveraging scarce resources (F-5). In addition, partnerships may have significant value in negotiating compromises or in building political support. These partnerships should be identified early in the process (L-5) because they require a solid understanding of watershed stakeholder interests.

Elected officials and regulators will often oppose an action that is also opposed by an organized, vocal stakeholder group. Therefore, it is very important that UAA sponsors identify and engage key regulators and elected officials that may be decision makers on the UAA or key individuals that influence the outcome.

Political dynamics tend to give disproportionate power to those that are organized and vocal, either for or against a UAA. For elected officials and regulators to allow the process to move forward they must understand the precepts behind the UAA so they can counter any allegations that will prevent the UAA process from moving forward. Remember that stakeholder opinion usually drives elected official views and can influence regulator reactions.

4.4.1.2 Develop a Comprehensive Work Plan and Strategy

Developing an effective UAA work plan and strategy at the beginning involves several elements including:

- Understanding the issues from all agencies involved, including state and federal fish and wildlife agencies for aquatic life UAAs;
- Agreeing on an approach;
- Integrating the UAA process with other regulatory programs; and
- Engaging U.S. EPA regional staff at the beginning of the process.

Each element is described below.

In addition to understanding the regulatory structure of the state standards and how they are implemented, stakeholders need to understand which regulatory agency and/or department is responsible for standards development and implementation and what issues are important to them. In some cases, this may be a different agency or department than that responsible for TMDL implementation, stormwater management, and/or NPDES permitting. If this is the case, all of these parties should be integrated into the process early so that key and potentially divergent issues and perspectives are understood at the outset of the UAA. It is important to know the decision-makers at the agencies, both water quality and fish/wildlife agencies, and understand their issues and drivers at the beginning of the UAA process (L-3, R-3) so that, if possible, the UAA is a win-win for all involved. This approach greatly increases the chances of success because all parties involved can support and provide constructive input into the UAA work plan and process. This step also helps to build trust between the regulatory agencies and stakeholders.

By identifying relevant agencies, stakeholders, and issues at the start, the primary goal of team integration from the beginning is to develop a mutually agreed upon work plan and strategy that meets all regulatory and stakeholder issues (S&T-1, P-2). This strategy should rely on identifying and addressing all potential challenges and future issues (L-3) in the work plan with an understanding of how the results would fit into the decision process. These challenges and issues, if not anticipated, can derail a UAA process. But, if they are identified early enough and a mutually agreed upon strategy is developed to address these issues, the likelihood of a successful resolution increases.

Standards changes require a great deal of planning and implementation time, typically on the order of two to three years, and the process is often complicated and delayed by unexpected issues related to other regulatory programs. Understanding and planning for issues related to TMDL schedules, NPDES permitting compliance timelines, and other regulatory timing drivers (L-1) is necessary. In particular, the UAA should be integrated with any relevant TMDL process (S&T-6). Ideally, the UAA should be completed before the initiation of a TMDL so that the right designated uses form the basis of the TMDL and subsequent NPDES permit limits. The National Academy of Sciences (2001) recommendations stated this clearly: "to ensure that designated uses are appropriate, use attainability analysis should be considered for all waterbodies before a TMDL is developed."

Last, but not least, in the strategy, U.S. EPA regional needs must be consulted. Depending on the region and which regulatory agency has authority over compliance areas such as TMDLs, NPDES permits, and CSO programs, identifying U.S. EPA regional issues early can be critical. Similar to recommendations L-3 and R-3 above, the extent to which U.S. EPA's issues are understood and integrated into the strategic plan can lead to a more successful resolution of the UAA process.

4.4.1.3 Develop a Decision Pathway

Developing a decision pathway at the beginning involves a complex integration of technical, regulatory, and political considerations. In particular, defining the critical terms that may not be fully defined in applicable state or federal regulations is important. Furthermore, all agencies should be engaged, using their guidance if available, to define a clear scope and decide on the use of models. All impacted parties, including downstream users, should be considered.

The strategic plan should include an appropriate "decision pathway" that incorporates a well-thought out technical scope and an agreement on the important issues (S&T-2, S&T-3, L-2, and L-6). An illustration of this pathway and the related questions is as follows:

- Identify the stakeholders and their issues.
  - Who needs to be involved, at what level of involvement, and what are their issues?

Have downstream users and their issues been considered?

- Define what is at risk and the urgency.
  - What are the cost and regulatory consequences?
  - What are the regulatory requirements and timetables?
  - What needs to be communicated to stakeholders?
- Define what is known about the watershed and waterbody.
  - What are the existing data?

What are the, standards, existing uses and desired uses?

What do the data tell you about existing conditions, stressors, and ability to meet those standards and uses?

- Define the candidate regulatory 131.10(g) UAA factors relevant to a potential use change.
  - What are the critical terms that need to be defined in context the factors and issues specific to the watershed?
- Define the scope of the UAA evaluations.
  - Are there state or federal guidance and/or protocols that are applicable to the UAA under consideration?
  - What data, analysis and/or modeling are needed to evaluate the applicable factors?
  - Are additional data, analysis or modeling needed?

How will the data, analysis and/or modeling be interpreted and used to evaluate the applicable factors?

These issues might appear to be deceptively simple; however, in practice, these are often the issues that prevent a UAA from being successful. Additional commentary is provided below on selected key elements.

One area that needs to absolutely be addressed is the definition of critical terms that may not be fully defined in applicable state or federal regulations (S&T-2, L-2b). These terms typically include "natural," "background," "feasible," "nonremedial," and "irreversible." Decisions on a UAA justification often hinge on the interpretation of definitions for these terms.

Without clear, mutually agreed upon definitions, these terms can greatly affect the interpretation of available data and cause disagreement between regulators and the regulated community.

Properly defining these terms will also help mitigate disagreement on how the data and modeling output will be interpreted and used, which is probably the most common difference of opinion in the UAA process. Consensus on this issue does not, however, mean that the UAA outcome should be predetermined before compiling or developing all of the data. Rather, this recommendation focuses on the need for a clear pathway for what data are needed and how the data will be interpreted and used in making decisions or judgments (S&T-3). Understandably, regulatory staff may want to wait until after they see the information to determine how it should be interpreted. However, these issues tend to be easier to resolve objectively before the data are available because the discussions can be theoretical. Helpful questions to ask and answer at the beginning might include, as *illustrations*:

- How much physical habitat is required to support a specific aquatic ecosystem?
- Are recreational uses applicable during or shortly after storms and, if not, over what post-storm duration?
- If the aquatic community is predominantly comprised of adults with limited juveniles, with an agreement at the beginning as to what constitutes "predominantly" and "limited," which standards or targets apply?
- What is the appropriate model, what are the model limitations, and how do these limitations affect the ability to make decisions on what is attainable?

Another strategy to help mitigate disagreement is to rely on available guidance. Federal guidance on UAAs is somewhat limited and state guidance varies greatly. Typically, no state guidance exists yet many states are in the process of developing their own guidance. In these circumstances, the UAA team, including both regulators and the regulated community, should work together to select guidance used in other states and/or to answer the questions relevant to a clearly defined decision pathway (R-2).

A well-defined decision pathway also has the added benefit of being a tool for developing a tighter technical scope and budget (F-4), and the necessary nontechnical elements of a UAA (such as public awareness and involvement, legal and regulatory requirements), regardless of who leads the UAA. This includes simple situations that may be associated with a less rigorous decision pathway, as well as complex situations that require a more detailed and better thought out decision pathway. In complex situations, for example, models are often required (S&T-4). Incorporating the modeling needs into the scoping process at the beginning of the UAA effort leads to a more realistic assessment of the effort required to complete the UAA (F-4). In addition, criteria for model "approval" by the various state and federal agencies should be determined before investing resources in modeling efforts.

To reiterate, investing resources into either answering the difficult questions at the beginning, or determining what information is needed to be able to answer those questions as part of the UAA, is a good investment. Although it might seem like a difficult initial challenge, the majority of successful UAAs have the hallmark of close early cooperation on these issues so that all parties' objectives and needs can be met in the process. If consensus cannot be reached at the beginning on many of these issues, it is better to understand these disagreements and re-assess the probability of a successful UAA before resources are unwisely invested in the process.

Last, when considering the issues and stakeholders, many UAAs focus on stakeholders that come to the table and the issues they bring. However, it is critically important to seek out issues and stakeholders and, in particular, to consider downstream uses (L-6) that might not initially be engaged but later could become a major obstacle.

Section 4.8 presents two illustrations of how the development of a decision pathway might be considered in hypothetical examples.

4.4.2 Conducting the UAA

Once a decision pathway has been defined and the UAA proponents want to proceed, two elements contribute to the successful completion of a UAA:

- Rely on sound science, good documentation and an inclusive stakeholder information and outreach process to gain approval; and
- Rely on a combination of technical and economic analyses to define "feasible."

4.4.2.1 Work to Gain Approval (Base the UAA on Sound Technical Analyses and Inclusive Stakeholder Involvement)

A successful UAA relies on many factors. However, without a sound technical basis that relies on good science and a public process that minimizes conflict, the other UAA elements do not matter. Each UAA situation is unique and there is no cookbook for how to use available biological, chemical, and physical data to determine existing and attainable uses. Given that fact, a well-defined decision pathway that lays out how much data and modeling are needed, acceptance criteria for the models, and how that data and modeling can and should be interpreted can help minimize disagreement (L-4). In short, any time that a UAA can be based on objective interpretation of the data and modeling versus subjective interpretation of the data, the stronger the UAA scientific and technical basis will be.

Coupled with sound science and thoughtful analysis is the need to maintain good documentation of references, analyses, data sources, modeling calibration/results, conversations, meetings, etc. This provides an administrative record of not only the sound technical basis that is the underpinning of a defensible UAA, but that the appropriate regulatory steps were followed in a timely fashion (L-2). A successful outcome, including avoiding litigation, requires two things: sound technical arguments and absolute compliance with regulatory rules and schedules. The administrative record is the measure for meeting these two needs.

4.4.2.2 Develop Thorough and Appropriate Economic Analyses

The issue of feasibility in a UAA analysis is best addressed with both technical and economic analyses. Determining what is "feasible" is an explicit component of UAAs and is described in 131.10(g) as follows:

"States may remove a designated use which is not an existing use, as defined in 131.3, or establish a subcategory of a use, if the state can demonstrate that attaining the designated use is not feasible because..." and then lists the six factors.

UAA Factors 1 though 5 are usually thought of as "technical" factors and, in fact, U.S. EPA has indicated that economic analyses are most appropriate for Factor 6 rather than, for example, technical Factor 4 regarding the operation of dams (U.S. EPA, 1998). Thus, it is evident that technical analyses alone can suffice for a legal determination for Factors 1 through 5. Nonetheless, some successful UAAs have also included economic analyses to reinforce the decision to modify designated uses even though Factor 6 may not have been cited as a legal basis for the decision. Useful economic analyses could include cost-benefit and cost-effectiveness analyses, which assist in stakeholder understanding and acceptance of what may be realistic or feasible regarding water quality improvement and attainability of standards.

As noted earlier, economic Factor 6 (affordability) has not been relied upon for many UAAs to date. If Factor 6 is to provide legal justification for a designated use change it may also be advisable that the analysis considers other indicators of affordability rather than restrict the analysis to what is contained in current U.S. EPA economic guidance. One such indicator is the economic and social impact on lower income households in the affected area, rather than relying solely or primarily on impacts to median income households.

The available economic guidance that is most commonly referenced is the U.S. EPA guidance on affordability (U.S. EPA, 1995). This guidance relies on a financial threshold for what is deemed to be substantial impact, and not on a marginal cost-benefit or cost-effectiveness analysis. The result is that the vast majority of UAAs cannot pass the U.S. EPA test for economic burden. Understandably, the U.S. EPA guidance presents a conservative economic threshold; however, this threshold does not reflect the economic realities that many in the regulated community face in complying with water quality standards. For example, in wastewater treatment there is a point where increasing marginal costs far outstrip marginal water quality benefits from more stringent effluent limitations. The U.S. EPA guidance is silent on this issue, even though public works agencies must justify their expenses to their local management and to the citizens they serve. Given these realities, there is certainly room for the development of additional economic guidance for UAAs (F-2).

The other area where UAAs can better incorporate financial considerations is to consider whether developing an economic value of a current or attainable use is needed (F-3). The economic value of the use refers to the idea that a waterbody has a value related to its use (fishable and swimmable). If the use were refined, how would that value change? For example, what would the economic effect be associated with moving from a fully supporting recreational use that includes swimming to a more limited recreational use that includes only boating? Again, U.S. EPA guidance is silent on this issue and the research team did not identify any case studies in which this has been done, though from a nontechnical standpoint, it may have a strong influence on the political and public perception of the UAA process.

4.4.3 Involving Stakeholders

As mentioned in numerous places in this guidance, including Sections 4.3 and 4.4.1.1, involving the full breadth of stakeholders early on and throughout the process is essential for success. This is needed not only to gain their approval, but also to identify key issues and ensure that the UAA study adequately addresses all issues.

On the issue of involving stakeholders, two basic recommendations were identified that might be neglected: fully engaging the public; and educating politicians.

4.4.3.1 Fully Engaging the Public

Independent of who leads the UAA, public involvement is important to a successful outcome (P-1), not just the engagement of regulatory and regulated stakeholders. Successful public involvement almost always entails more than the minimum public notification requirements spelled out by rulemaking processes. Important goals of the public involvement component of a UAA include building trust, increasing understanding of the UAA process, soliciting feedback, and learning more about the waterbody and available data resources. Merely relying on public notice and/or one hearing to obtain feedback is a disaster waiting to happen. Invariably, mistrust and strong objections may result, many of which are not easily countered.

As with any public awareness campaign, face-to-face communication is the most effective method because questions can be asked and, hopefully, answered before the issues escalate (P-4). In addition, this forum can help build trust between UAA proponents and the public, many of who may not fully understand what a UAA is or what the outcomes of a UAA process may be (P-3). Furthermore, this is the best opportunity to inform the public of the large economic investments at stake and to engage them in the discussion of the various consequences and how much they are willing to pay for various outcomes (P-6).

In general, there is not a one-size-fits-all process for engaging the public. Designing a public education and outreach process has to be tailored to the specific stakeholders involved and the issues at hand. A range of approaches can be employed including, as examples:

- distribution and mailings of pamphlets, brochures and newsletters;
- news releases and articles;
- web pages, online discussion groups and other electronic communications;
- public hearings and public meetings;
- workshops and seminars;
- discussion groups;
- library repositories of information and reports; and
- individual meetings, phone calls and visits.

The only general rule to public education and outreach is that the more frequent and personal the process, the more effective is the engagement. The UAA sponsors need to artfully assess the needs of the stakeholders and design an effective yet adaptable process.

4.4.3.2 Educating Decision Makers

There are certainly different elements of the public that may require or ask for education on the UAA process. Decision makers, including elected officials and watershed advisory groups, might need special focus so that once a UAA is completed decisions can actually be made (P-5). Often, these groups are especially interested in, and sometimes responsible for, economic impacts. These impacts need to be clearly defined in terms that officials can readily understand (P-5).

It is critically important that these elected officials, decision-makers and key political figures be well informed and educated throughout the conduct of the UAA. By routine education their implicit, if not explicit, support may be attained. Unexpected surprises to them can only mean trouble for the UAA process and often considerable effort to remedy distrust. Therefore, a little extra attention in this area is effort well spent.

4.4.4 Resolution

An underlying theme to this guidance is that a successful UAA outcome can be defined broadly, where alternative regulatory solutions result from consideration of a UAA, or where a UAA leads to widespread public and political recognition that current uses are appropriate.

Getting there requires a great deal of planning and discussion at the beginning of the process, as well as a commitment throughout the process to base the UAA on sound science, appropriate economic considerations, and public input. Even if all of the recommendations contained herein were followed, a UAA outcome might still be different than what was originally anticipated. Going back and reviewing the parties' objectives and needs is a good idea because it represents being open to an alternative regulatory solution (R-5). It is important to not be wed to one solution (i.e., a specific change in the use), but fully consider what *practical* issues are at stake, and determine if alternative regulatory solutions solve those practical challenges and meet the concerns of the full range of stakeholders.

Examples of alternative solutions are many but could include a temporary variance, a long-term compliance schedule with milestones for reconsideration, a use refinement versus a removal, site-specific water quality criteria, a long-term consent order, or additional evaluations to support regulatory analyses or requirements. These and other alternatives are also worthy of consideration (see Chapter 3.0).

Findings from the research support the idea that there are many good lessons learned that can be applied by anyone considering or pursuing a UAA. It is not unusual to end up at a different regulatory solution than what was envisioned at the beginning of the process.

4.5 Hypothetical Examples of Decision Pathways

Two hypothetical examples are presented below that illustrate the development of a decision pathway for a UAA, as discussed in Section 4.4.1.1. The examples highlight the UAA process as applied in an urban stream with CSOs and a reservoir suffering from eutrophication. Development of such a decision pathway helps in identifying stakeholders and key issues early in the process, defining data and modeling needs, and providing a general framework for conducting the UAA and integrating the diverse and potentially complex issues.

Hypothetical Example #1 – Urban Stream with CSOs.

Example Description	♦ Intermittent but flashy urban stream. ♦ Concrete lined, limited-access channel. ♦ Essential for flood control. ♦ Receives CSO discharges from an urban business district. ♦ Designated for primary contact recreation; bacterial criteria are exceeded during storms. ♦ Downstream of the city the stream becomes natural, rural and perennial; bacteria criteria also exceeded during storms; stream eventually discharges to a large river where bacteria criteria are met.
Define the Stakeholders & Issues	♦ City residents and politicians are concerned with cost as it relates to water rates, and use of the stream. ♦ NGOs are concerned with uses and ecology. ♦ Regulators are concerned with compliance to legal requirements. ♦ Downstream users have uncertain interests. ♦ ***Agree on the stakeholders and scope of involvement.***
Define the Risk and Urgency	♦ Size and operational characteristics of CSO control facilities, and hence cost, stipulated in a required LTCP are strongly dependent on how much overflow is allowable and for how many events. This, in turn, is dependent on the designated use and applicable criteria. ♦ CSO LTCP is proposing a reduction in overflows to only several times per year during only very large storm events, but still at a very large cost. ♦ CSO LTCP is being prepared and must be completed in 2 years; a compliance order has a 10-year schedule for facilities. ♦ To meet schedule, engineering and construction activities need to be started soon, thus UAA and standards decision need to be resolved as soon as possible. ♦ ***Agree on a schedule.***
Define What is Known About the Watershed & Waterbody	♦ Land Use: urbanized area transitioning to rural. ♦ Stream Channel: concrete-lined transitioning to natural. ♦ Hydrology: intermittent and flashy transitioning to perennial. ♦ Water Quality: violation of bacteria standards during and after storms both in the urban areas and downstream. ♦ Pollutant Sources: both urban and rural. ♦ ***Educate and agree on the issues.***
Define Candidate 131.10(g) Factors	♦ Could potentially apply Factor 2 (natural flow conditions), Factor 3 (human-caused conditions), Factor 4 (hydrologic modifications), and possibly Factor 6 (socioeconomic impact). ♦ Factor 4 (hydro modifications that are essential for flood protection) could be sufficient, but needs to be confirmed. ♦ ***Agree on best factor for UAA analysis.*** ♦ Key terms to define are "natural," "not feasible to restore," and "cannot be remedied." ♦ ***Agree on definitions.***

Define Scope of UAA Evaluations	♦ Existing sewer model can be used to quantify frequency duration and magnitude of overflows. ♦ Existing stream model can be used to simulate stream response and compliance to standards. ♦ Need model of nonpoint loads from rural sources. ♦ ***Agree on need for new data and/or models.*** ♦ ***Agree on model applicability and credibility.*** ♦ Need information on recreational uses in the urban and rural areas of the stream. ♦ Need information on stream velocities during storms. ♦ ***Agree on current uses.*** ♦ Data and models can be applied to predict bacteria levels, stream depths and velocities, and recreational use patterns in areas with potential recreation use, examined over a typical year. ♦ Results can be used to evaluate appropriateness of a temporary high flow exemption to recreation uses and where and when the exemption would apply. ♦ ***Agree on how models will be used.*** ♦ UAA will have to establish that primary contact recreation would be dangerous during large events remaining after CSO controls are built; and what is the appropriate duration of the temporary suspension, and at what point downstream. ♦ ***Agree on criteria for determining safety.*** ♦ Engineering and cost evaluations will need to be conducted to evaluate the cost implications of more restrictive and expensive controls.
Decision	♦ ***Agree on forum, participants, and criteria for final decision making.***

Hypothetical Example #2 – Reservoir with Eutrophication.

Example Description	♦ Large hydroelectric reservoir, constructed in 1950s; undergoing federal relicensing. ♦ Point and nonpoint sources contribute nutrients upstream of reservoir. ♦ Designated use is cold water aquatic life. ♦ DO criteria are not met in the hypolimnion of the reservoir. ♦ Downstream of the reservoir, in the free-flowing river, coldwater uses and DO criteria are met.
Define the Stakeholders & Issues	♦ Hydroelectric facility wants continued operation to maximize power generation. ♦ Upstream municipalities do not want unfair nutrient limits imposed on their effluent; also concerned about being pushed on stormwater permit conditions. ♦ Agricultural producers in the watershed don't want conditions put on their practices or productive land taken out of production. ♦ Regulators are concerned with compliance to legal requirements, including development of a TMDL within 3 years. ♦ Upstream and reservoir users want trout fisheries and water levels maintained. ♦ Downstream users and landowners want cold water uses and existing flows/flood control to be maintained. ♦ ***Agree on the stakeholders and scope of involvement.***
Define the Risk and Urgency	♦ Re-licensing required for hydroelectric facility to continue operations, including water quality certification conditions in the new license to meet current DO criteria in the reservoir. ♦ A TMDL is scheduled to be completed within 3 years; a TMDL to meet current water quality criteria for DO in the reservoir is expected to be stringent and costly to meet. ♦ NPDES permit holders awaiting permit renewal and need to consider potential treatment options in long-term facility planning. ♦ A UAA could be completed and new standards adopted and approved before TMDL is completed. ♦ ***Agree on a schedule.***
Define What is Known About the Watershed & Waterbody	♦ Hydrology: effluent flows into reservoir are significant and, therefore, hydraulic residence time and power generation are dependent on effluent flows. ♦ Water Quality: violation of DO criteria in the hypolimnion of the reservoir, typically during late summer. ♦ Fisheries: Trout use the reservoir only in winter and only for rearing when it is colder and DO concentrations are higher and meet current standards. ♦ Pollutant Sources: large contributing watershed with both municipal point sources and urban and rural nonpoint sources. ♦ ***Educate and Agree on the issues.***
Define Candidate 131.10(g) Factors	♦ Could potentially apply Factor 3 (human-caused conditions) and/or Factor 4 (hydrologic modifications), and possibly Factor 6 (socioeconomic impact). ♦ Factor 4 (hydrological modifications that makes river susceptible to eutrophication) could be sufficient but needs to be confirmed. ♦ ***Agree on best factor for UAA analysis.*** ♦ Key terms to define are "existing use," "feasible to restore," and "cannot be remedied." ♦ ***Agree on definitions.***

Define Scope of UAA Evaluations	♦ Historical data (back to 1975) available in university archives and fishery agency reports. ♦ Relicensing process provides extensive water quality and aquatic biota data. ♦ Validated and agency accepted water quality model for the upstream river and reservoir are available. ♦ Data and modeling of nonpoint sources in watershed are not available. ♦ Concerns about changes to downstream water quality may drive need for data and modeling downstream of reservoir. ♦ ***Agree on need for new data and/or models.*** ♦ ***Agree on model applicability and credibility.*** ♦ Evaluate available data to confirm present use of reservoir by trout: seasonal and spatial use. ♦ Evaluate operation of hydroelectric facility and how power generation is related to upstream flows. ♦ Confirm downstream cold water fisheries. ♦ ***Agree on current uses*** ♦ Apply models to evaluate the attainability of current DO criteria for a range of upstream point and nonpoint source controls, including effluent reuse. ♦ Reservoir aeration could be part of the solution and can be evaluated with the model. ♦ Effluent reuse would reduce hydraulic retention time in the reservoir, which could lead to water quality impacts downstream; apply models to determine potential impact. ♦ Model results combined with fisheries data could be used to develop subcategories and/or seasonal uses of the reservoir by trout, and possibly define attainable reservoir-specific DO criteria for the hypolimnion. ♦ ***Agree on how models will be used.***
Decision	♦ ***Agree on forum, participants, and criteria for final decision making.***

CHAPTER 5.0

HIGHLIGHTED CASE STUDIES

Six UAA efforts were selected from the 54 UAAs researched for presentation as case studies. These case studies illustrate the key challenges faced in the UAA process and highlight how best to address those challenges, consistent with the findings and recommendations presented previously in this report. The case studies were selected to represent three common and challenging situations where UAAs are being considered nationwide: wet-weather impacts; urban settings; and effluent dependent or dominated streams. The case studies presented are as follows:

- Wet-weather impacts
  - Boston Harbor CSOs
  - California engineered flood channels
- Urban settings
  - Chesapeake Bay
  - Cuyahoga River ship channel
- Effluent-dependent or -dominated streams
  - Spokane River
  - Lower French Gulch and Blue River

To highlight the connections between these case studies and the findings of this study, Finding labels have been included in parentheses in appropriate places that match those used in Chapters 3.0 and 4.0 for easy reference. These case study summaries are intentionally brief to only illustrate the UAA process and the findings. For further information on each case study, the reader is referred to the references cited in the text, and Appendix B which provides contact information for professional(s) with knowledge of the UAA.

5.1 Boston Harbor Combined Sewer Overflows

The Greater Boston, Massachusetts Area (Figure 5-1) has a very old combined sewer and storm water system, with many CSOs. Over the past two decades, the Massachusetts Water Resources Authority (MWRA) has been working on eliminating many of these CSOs, yet it was found that eliminating all CSOs would result in "substantial and widespread economic and social impact." Using the regulatory options outlined in Massachusetts Department of Environmental Protection (MA DEP) policy for CSOs, the MA DEP and the MWRA assigned areas of the Boston Harbor with a CSO impact designation through a UAA (L-5).

5.1.1 Background

Like many older cities in the United States, the City of Boston and its surrounding communities have a combined storm and wastewater system. When capacity within the wastewater collection system is exceeded during a storm event, CSOs result and violations of water quality criteria may occur. The MWRA has been working towards treating or eliminating most of the CSOs. Their effort is a result of a court order originating in 1985 (the "Boston

Harbor Case") in which the MWRA was required to significantly upgrade their wastewater system. A 1987 stipulation to the Boston Harbor Case required that MWRA also be responsible for developing a control plan to address CSO discharges.

Over the course of the next 10 years, the MWRA developed a plan that outlined the elimination, mitigation, or treatment of many of the CSOs. This was consistent with U.S. EPA's Combined Sewer Overflow Control Policy (U.S. EPA, 1994), which requires that CSO dischargers develop LTCPs that outline how they plan to ultimately protect receiving water designated uses. When elimination of the CSOs or their impacts is not feasible, the Policy recommends pursuing a UAA with the possibility of having refined or partial use categories or a variance. The CSO Policy encourages states to establish a means by which this may be accomplished. In 1997, Massachusetts developed a document that outlines the regulatory options for CSOs (MA DEP, 1997) (R-1).

In summary, the MA DEP, working together with the MWRA, developed a UAA based on the LTCP. The CSO abatement plan included measures to eliminate CSO discharges to sensitive use areas and to minimize or treat CSO discharges to other receiving waters to the maximum extent feasible. The basis for the UAA was "substantial and widespread economic and social impact" [40 *CFR* 131.10(g)(6)] (F-1 Exception). The result of the UAA was a use change (CSO designation) in the Inner Boston Harbor (December 31, 1997) and a variance for the Charles River (October 1, 1998). Other areas were also considered, but only the Inner Boston Harbor and the Charles River will be highlighted in this case study. The following describes the process for which Boston developed and achieved their UAA results.

5.1.2 Conducting the UAA

The goal of MWRA's LTCP was to develop a plan for elimination of all of their CSOs. Consequently, MWRA collected data and modeled their system using predictive models such as Storm Water Management Model (also later utilizing InfoWorks) for modeling the sewer system and Modeling of Urban Sewer Environment, or MOUSE, for water quality modeling (S&T 4). The USGS also conducted special studies that evaluated the influence of stormwater on the system. Extensive water quality data and information were gathered and are still being gathered to continuously characterize the system as changes are made. Finally, the control measures to eliminate each CSO and the costs to implement these controls were identified (S&T-5).

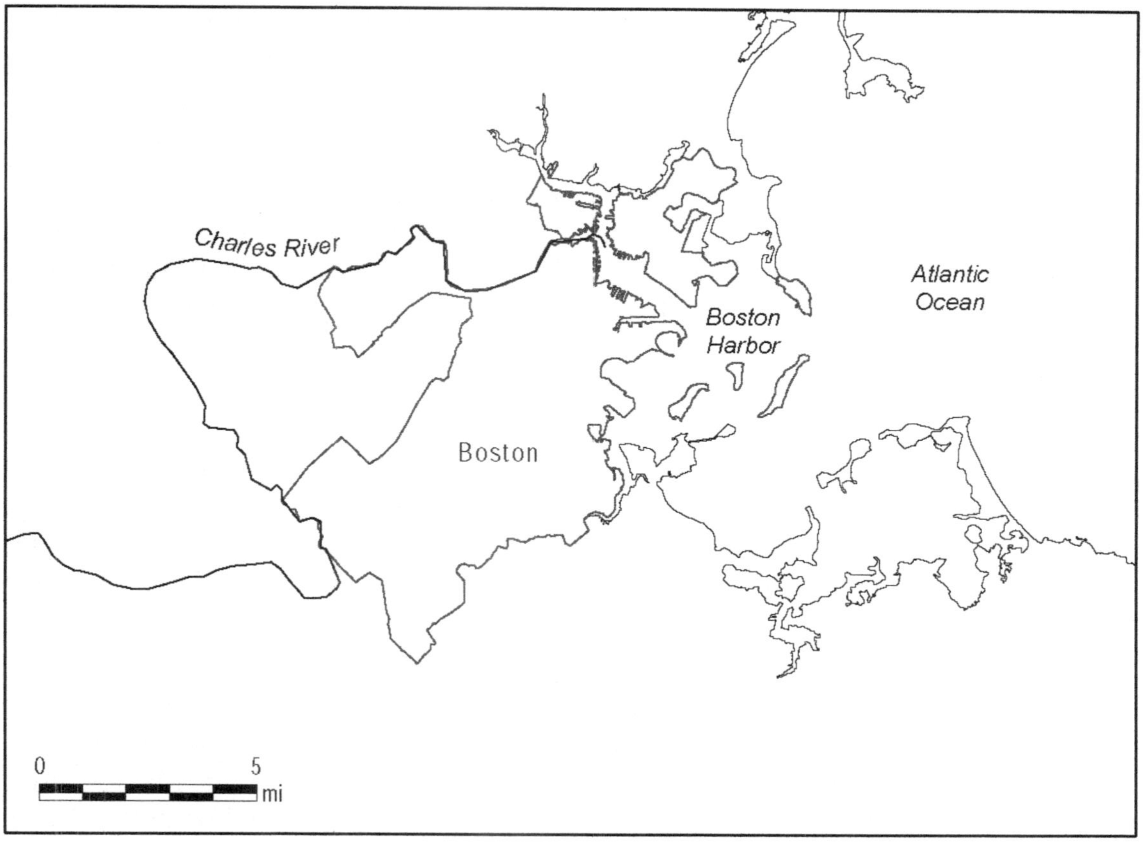

Figure 5-1. Map of Boston Harbor Area

When it became clear that the costs to eliminate these CSOs exceeded $1 billion dollars and would place a significant burden on rate payers, MA DEP reviewed their regulatory options as outlined in the MA DEP Policy (PA-6).

To describe the options outlined in the MA DEP policy, one must first be familiar with the water quality designations in Massachusetts. In Massachusetts, waterbodies are classified as either A, B, or C (SA, SB, or SC for marine waters). All waters across the state are currently classified as either Class A, source of public water supply and "fishable/swimmable" (no CSOs allowed into drinking water supplies), or Class B, "fishable/swimmable." The regulatory options for CSOs outlined in the MA DEP Policy include:

- Class B or SB – CSOs are eliminated;
- Class B (CSO) – CSOs remain but must be compatible with water quality goals;
- Variance – CSOs remain when allowed under a short-term modification of water quality standards through an NPDES/MA permit;
- Partial Use Designation – CSOs remain with moderate impacts resulting in intermittent impairment of water quality goals; and
- Class C – CSOs remain, causing permanent and sustained impairment so that Class B water quality goals cannot be met.

To change the designation of Boston Harbor from a Class SB to a Class B (CSO) or any other appropriate designation, a UAA had to be conducted. Economic hardship and cost-effectiveness were the drivers behind this UAA. MA DEP utilized the detailed cost-effectiveness information included in the MWRA LTCP and a subsequent MWRA affordability analysis (F-2). Ultimately, U.S. EPA wanted to see the economic analysis be conducted based on the *U.S. EPA Economic Guidance for Water Quality Standards* (US EPA, 1995). Although MWRA's approach and the U.S. EPA approach have many things in common, there were a few differences including the fact that MWRA's approach considered cost-benefit, specific water and sewer issues, and local social conditions, whereas the U.S. EPA guidance only considers affordability. These differences were notable in the Boston area because such things as housing costs in Boston are much higher than the national average, which the U.S. EPA guidance did not consider.

5.1.3 Resolution

U.S. EPA New England was responsible for final approval of the UAA. U.S. EPA's response to the UAA was that they agreed with the findings, yet disagreed with the methodology (dismissing the state's inclusion of cost-benefit as an important consideration). U.S. EPA agreed with MA DEP's findings because, after U.S. EPA conducted the economic analysis using the U.S. EPA guidance, the finding was still that economic hardship would exist if all CSOs were to be eliminated.

The UAA process was successful due, in part, to the public outreach and education conducted by the MWRA (PA-2). In addition to the formal public comment and review process outlined in the Massachusetts Environmental Protection Act (MEPA), the MWRA held neighborhood workshops to educate the community on the issues (PA-1). The MWRA also reached out to watershed and advocacy groups (PA-4). Additionally, a well-organized advisory board was useful in keeping the focus on spending the money on cost-effective projects.

The successful outcome was because of the commitment of both the MWRA and the MA DEP, and the effort put forth for the development of supporting materials (i.e., LTCP), and the public outreach and education. The Inner Boston Harbor was changed from a Class SB to a Class SB (CSO). The Lower Charles River and the Mystic River watershed were granted CSO variances (R-5). Not enough information was available to support a permanent change, and initial findings identified stormwater to be a larger pollutant load than CSOs. With the variances, the MWRA will be allowed to conduct further sampling and analysis of storm drains "to determine the relative merits of additional controls over stormwater as compared to additional CSO controls."

5.2 California Engineered Flood Channels

In July 2003, the California RWQCB - Los Angeles Region (Los Angeles RWQCB) adopted a change in its Water Quality Control Plan to suspend recreational designated uses in specified engineered channels (Figure 5-2) during specified wet-weather conditions (R-5). The Los Angeles RWQCB proposed to suspend the recreational designated uses in engineered flood control channels where access can be restricted during and immediately following heavy storms to address the physically unsafe conditions in these channels. This UAA case study represents a situation in urban areas where waterways have been engineered to convey large volumes of storm water efficiently and, as a result, cannot safely be used for recreation during and immediately after heavy storms. Because of both the inherent design of "engineered flood channels" that creates life-threatening flow conditions during and immediately after heavy storms, and county policies restricting access during these same periods, the UAA concluded that recreational uses do not exist and are not fully attainable during defined storm conditions. The adoption of the UAA explicitly recognized that temporary suspension of recreational uses does not suspend the requirement to meet other designated uses. Therefore, it should not alter or interfere with dischargers' responsibilities and commitments under downstream beach TMDLs (i.e., the upstream use change does not relieve or diminish obligations to reduce bacteria loading at the downstream beaches) (L-6).

5.2.1 Background

Before the UAA, the issue of protecting the public during swift water conditions was already widely recognized by public safety agencies within the Los Angeles area. The Los Angeles County's Multi-Agency Swift Water Rescue Committee had existing protocols for restricting access to flood control channels and preparing for possible rescues during storm events. In the UAA process, the Los Angeles RWQCB strove to be consistent with local agencies' existing protocols for protecting public safety in flood control channels. The UAA was conducted by staff at the Los Angeles RWQCB following federal regulations pertaining to UAAs (40 *CFR* 131.10).

The UAA applies to waterbody segments that meet the definition of an "engineered flood channel," which is defined as inland, flowing surface waterbodies with a box, V-shaped, or trapezoidal configuration that have been lined on the sides and/or bottom with concrete. The explicit purpose of these channels is to reduce flooding in urbanized areas by conveying storm water runoff to the ocean or other discharge point as efficiently as possible.

All of the 39 waterbody segments are assigned a REC-1 (water contact recreation) year-round use, with the majority of waterbody segments also assigned a REC-2 (noncontact water recreation) year-round use. Although access to all of the waterbody segments is physically restricted or prohibited by the local county during storm events, the recreational use designations meet the "swimmable/fishable" goal of the Los Angeles RWQCP (Basin Plan) and the federal Clean Water Act. The UAA explored the temporary suspension of recreational uses during and immediately following specified storm conditions, specifically during and 24 hours after a storm with rainfall of ½ inch or greater (R-5).

The Los Angeles RWQCB intended to use the UAA as the basis to modify their Basin Plan and incorporate temporary suspensions of recreational designated uses.

5.2.2 Conducting the UAA

The UAA relied on Factor 2 (natural, ephemeral, intermittent, or low-flow conditions prevent attainment of the use) and Factor 4 (dams, diversions, or other types of hydrological modification preclude the attainment of the use). Factor 2 reflected on the naturally flashy nature of local storm events, the original reason for constructing the engineered flood channels, whereas Factor 4 acknowledged that the inherent flashy nature of storms was exacerbated by hydrologic modifications in the subject channels. The UAA also recognized that it is currently not feasible to restore the waterbody segments to original conditions or operate modifications to attain the use during storm events (L-2, R-5).

Staff determined that the UAA should apply to a category of waterbody segments, those being the engineered flood channels. Staff identified 39 waterbody segments that both met the definition of an engineered flood channel and had prohibited access during defined wet-weather conditions. Staff conducted field surveys to confirm access restrictions and determine physical characteristics such as channel flow and geometry, and construction materials on the bottom and sides of the channel.

Staff also recommended an appropriate rainfall trigger for the temporary use suspension. The decision to rely on rainfall to define the trigger, versus waterbody specific characteristics such as velocity and flow, was based on a widespread network of rain gages. The rainfall trigger recommended by staff was consistent with existing protocols used by the Swift Water Rescue Committee to lock access gates and put rescue personnel on alert. Three storm alert levels had been defined by this committee based on storm warnings, and the UAA recommended that the lowest alert threshold be used (rainfall equal to or greater than ½ inch as measured at the nearest rain gage). In addition, the use suspension would apply for 24 hours following an event, consistent with the county policy to restrict access during this period. Finally, the rainfall trigger was also consistent with the fact that most of the historical swift water rescues occurred when rainfall was equal to or greater than ½ inch.

Importantly, the UAA recommended that suspension of swimmable recreational uses did not suspend other water quality objectives set to protect fishable recreational uses, aesthetic aspects, or aquatic life. Also, the UAA recommended that temporary suspension of recreational uses should not alter or interfere with dischargers' responsibilities and commitments under downstream beach TMDLs (that is, the proposed use change should not relieve or diminish discharger obligations to reduce bacteria loading at the beaches) (L-6).

Public involvement was consistent with routine basin plan amendment processes and included the following components: (1) a public workshop was held; (2) a draft staff report, tentative resolution, and basin plan amendment was released to the public for comment and agency response; (3) notice of hearing and notice of filing were published; and (4) multiple open board meetings addressing the project scope and outcome were held (PA-1, PA-2, PA-4, PA-6).

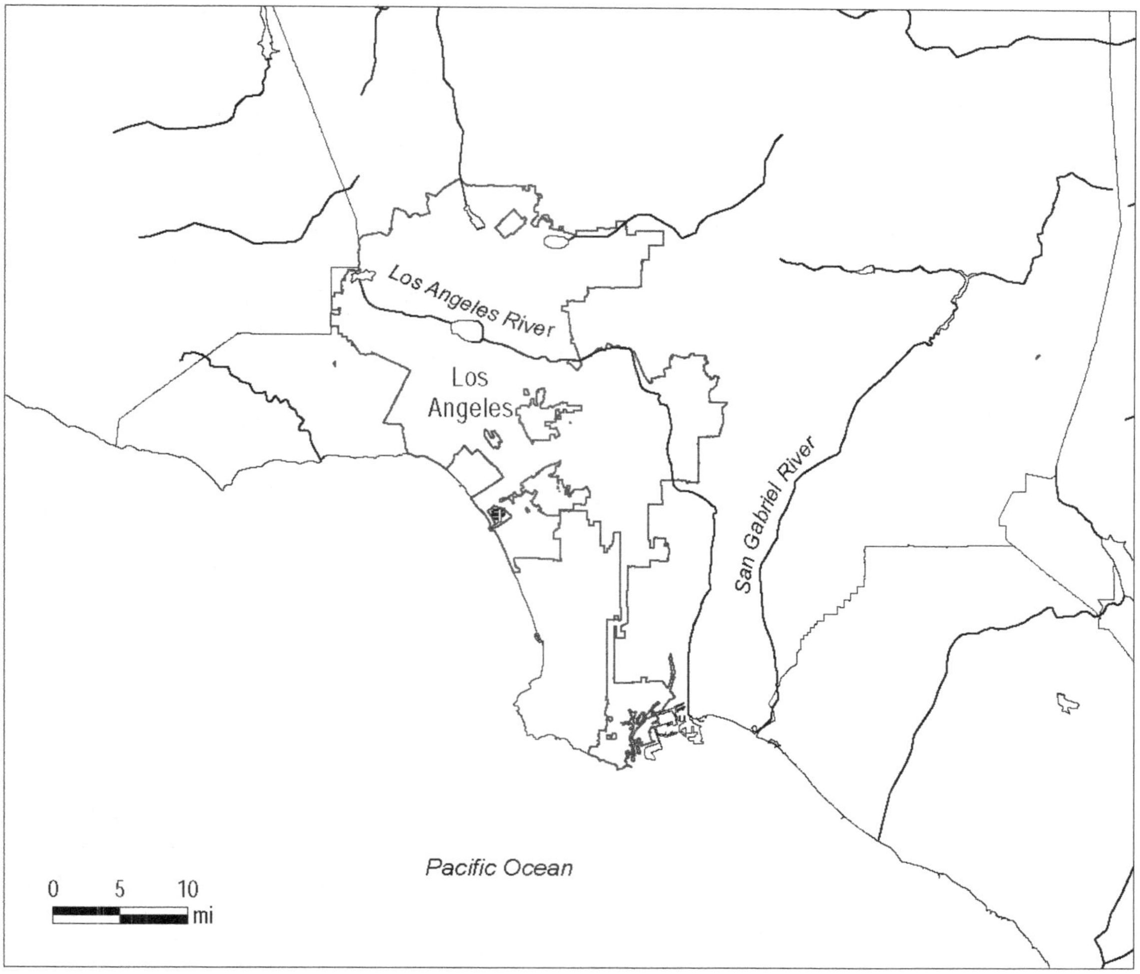

Figure 5-2. Map of Greater Los Angeles Region.

5.2.3 Resolution

In July 2003, the Los Angeles RWQCB approved the recommendations in the UAA and amended the Los Angeles Water Quality Control Plan (L-1) (Los Angeles RWQCB, 2003). Importantly, the amendment resolution simultaneously adopted the temporary use suspension and committed the Los Angeles RWQCB to: (1) re-evaluate the attainability of recreational uses in the future (within 3 years following the plan amendment); (2) support efforts to reclaim engineered channels as natural waterbodies; and (3) support the beneficial reuse of stormwater. A key factor in the success of this process is that it reflected a widely held belief, consistent with existing public safety policies, that swimming does not and should not occur in engineered flood channels following heavy storm events (R-6). In this sense, this UAA was fairly simple because the technical factors were straightforward relative to other UAA situations. Following approval by the State Board in November 2003, U.S. EPA Region 9 approved the amended Los Angeles Water Quality Control Plan in August 2004 (L-1).

5.3 Chesapeake Bay

The Chesapeake Bay UAA effort (Figure 5-3) was unique because of its geographical scale, comprehensive scope, and interagency cooperation (S&T-1). The effort addressed aquatic life issues for a 64,000-square-mile watershed, which is thought to be the largest UAA effort to date. The technical review in support of the UAA resulted from air, water, and biological monitoring and modeling (S&T-4). It also included a biological opinion by NOAA Fisheries (R-3) and an economic analysis. The effort was led by the Chesapeake Bay Program, a state/federal cooperative partnership between U.S. EPA Regions 2 and 3, U.S. EPA headquarters, six states, the District of Columbia, the Chesapeake Bay Commission, and two river basin commissions (S&T-1, L-5, L-6, and R-4).

Historical data and extensive monitoring and modeling (S&T-4) of the Chesapeake Bay revealed that Bay states' (Maryland and Virginia) water quality standards for DO were unattainable. Thus, the collaborators were focused on determining attainable water quality standards and the geographical extent of the designated uses to which the published Chesapeake Bay habitat-specific water quality criteria should apply. Several new subcategories of aquatic life use designations were developed and delineated within the Bay and its tidal tributaries and embayments, including the specific season(s) in which each of these uses occurs (R-5) (U.S. EPA).

5.3.1 Background

The Chesapeake Bay UAA was initiated in August 2000 after the signing of the *Chesapeake 2000* agreement. *Chesapeake 2000* was signed by the Chesapeake Bay Commission chair, the Governors of Maryland, Pennsylvania, and Virginia; the Mayor of the District of Columbia; and the U.S. EPA Administrator. In this agreement, parties committed to define water quality conditions necessary for protection of aquatic living resources and to make an effort to adopt new or revised water quality standards consistent with desired water quality goals. New York, Delaware, and West Virginia agreed to the same water quality commitments through a separate six-state memorandum of understanding with U.S. EPA. This set the stage for a comprehensive, basin-wide UAA conducted with a large consortium of agency support (S&T-1, R-4). The UAA effort was led by the Chesapeake Bay Program partners through its Water Quality Steering Committee (senior water quality program managers from all six states, the District of Columbia, U.S. EPA, the Chesapeake Bay Commission, and the two river basin commissions).

The Chesapeake Bay is the largest estuary in the United States and is home to more than 3,600 species of plants and animals and more than 16 million people. The Bay supports valuable commercial fisheries, is a shipping and commerce center, supports habitat for fish and wildlife, and provides recreational opportunities for residents and visitors. The Chesapeake Bay receives much of its water volume from the Atlantic Ocean, but also significant freshwater flows from the 64,000-square-mile watershed. Major rivers in the watershed include the Susquehanna, Patapsco, Choptank, Patuxent, Potomac, Rappahannock, York, and James, although there are more than 100,000 streams that flow into the Bay. The watershed lies within parts of six states: New York, Pennsylvania, West Virginia, Virginia, Maryland, and Delaware. The District of Columbia lies completely within the Chesapeake Bay watershed.

The Bay is approximately 200 miles long and varies in width from 3 to 35 miles. Including its tidal tributaries, the Chesapeake Bay has approximately 4,500 square miles in surface area and has 11,684 miles of shoreline (U.S. EPA, 2003a). The large watershed but relatively small surface area of water means that land use changes can dramatically affect the Bay. The Bay is also very shallow and, therefore, is very sensitive to temperature and mixing caused by wind and tidal effects.

Chesapeake Bay receives drainage from some heavily populated areas, with approximately 16 million people inhabiting the watershed (U.S. EPA, 2003a). Major industry in the area includes metallurgy, steam power, internal combustion engines, chemical engineering, and electronics (U.S. EPA, 2003a). Commercial fishing, forestry, and water-based recreation are also important contributors to the economy. Washington, D.C., the country's political headquarters, is also in the watershed. Land use in the watershed was reported to be 60% forest, 28% agriculture, 4% urban/suburban and 8% mixed open (Chesapeake Bay Program, 2006). Agriculture, urban and suburban development, and forestry activities have led to increased sediment and nutrient delivery to the Bay.

The pollutants considered in the Chesapeake Bay UAA were nutrients (nitrogen and phosphorus) and sediments, which were adversely affecting DO, water clarity, and chlorophyll *a* levels in the Bay. The designated use considered was aquatic life support. Because portions of the Bay are highly urbanized, pollutant sources included municipal wastewater treatment plants, combined and separate sanitary sewer overflows, urban stormwater, industrial discharges, and onsite sewage disposal systems. Up to a third of the nitrogen loads originate from atmospheric deposition to the watershed and tidal surface waters. The Bay also receives nutrient and sediment runoff from extensive agricultural lands throughout the watershed; natural biogeochemical processes and ocean exchange also supply some nutrients and sediments to the Bay.

The Chesapeake Bay UAA was pursued to determine if the then existing state water quality standards for DO were attainable. Stakeholders were concerned that future TMDLs focused on restoring the Bay's tidal waters would be developed based on potentially unattainable standards. The problems associated with land use in the watershed were impossible to mitigate without socioeconomic ramifications, and some evidence indicated that, historically (pre-European settlement), the Bay naturally received sufficient nutrient and sediment loads to result in low DO conditions in the deep waters of the Bay's mainstem. The basin-wide UAA analysis was conducted to provide use attainability-related information on a watershed-wide scale to help promote coordination and consistency across all seven jurisdictions. Also, all four jurisdictions with tidal waters (Maryland, Virginia, Delaware, and the District of Columbia) were required to calculate TMDLs for loads to the Bay for their jurisdictional waters by 2010. Therefore, the timeline for the UAA was set to have new designated uses in place before 2010 so that water quality standards would be met and the need for TMDLs potentially eliminated (S&T-6).

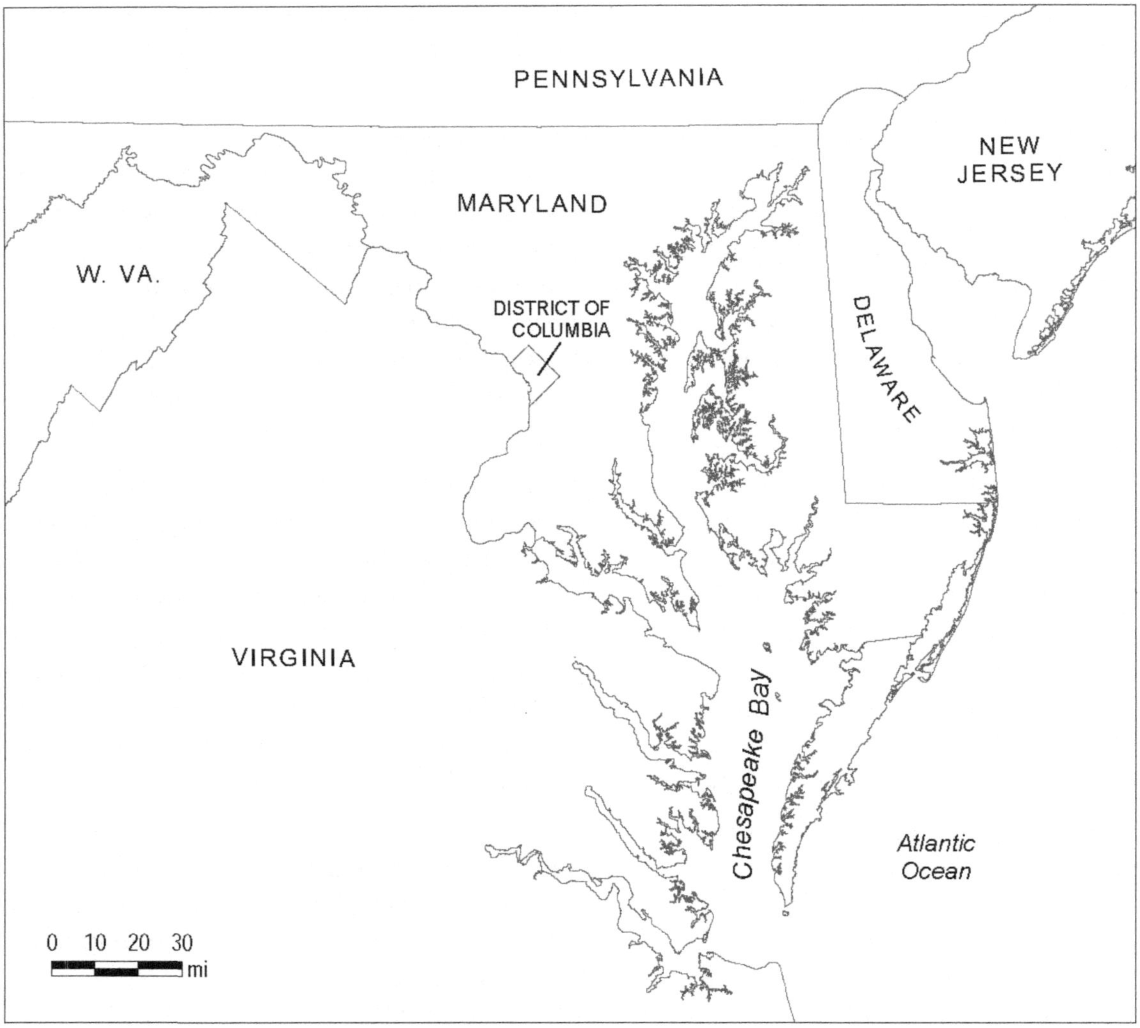

Figure 5-3. Chesapeake Bay UAA Study Area.

5.3.2 Conducting the UAA

The diverse consortium of stakeholders and agencies involved in the Chesapeake Bay UAA aided in the success of this large, multi-state UAA (L-5, L-6, PA-1, PA-2). U.S. EPA played a large part in the decision-making process and in developing the guidance for the UAA effort, in direct and constant coordination with the lead water quality, natural resource and agricultural agencies from each of the six states and the District of Columbia (L-1, R-4). U.S. EPA regional and headquarters' counsels were involved throughout the process to ensure that all legal and regulatory requirements were met (R-4). Representatives from the environmental advocacy and municipal wastewater communities were directly involved through the development and application of the Bay UAA.

The Chesapeake Bay UAA was largely completed without guidance directly applicable at the scale and complexity of the six-state watershed. Because of the scope of this effort, state guidance was not applicable. State rules were followed when necessary, but U.S. EPA was the lead agency and consulted their legal counsel throughout the entire process (R-4). NOAA Fisheries also developed a Biological Opinion because an endangered fish species was present in the Bay (R-3).

Extensive discussions about the process that should be used to establish designated uses and how to define natural and human-caused conditions occurred early in the process (S&T-1, S&T-3, L-4, and L-5). The involvement of U.S. EPA and state agency technical leads and legal counsel, as well as numerous other parties, ensured that the documents produced were consistent with expectations, and that regulatory terms were defined at the outset (L-5, R-4). This also facilitated smooth progress throughout the UAA effort.

The two primary 131.10(g) factors cited in the UAA in support of the recommended use refinements were: Factor 1 (naturally occurring pollutant concentrations prevent the attainment of the use) and Factor 3 (human-caused conditions prevent attainment of the use and cannot be remedied or would cause more environmental damage to correct than to leave in place). These two technical factors were supplemented with an economic assessment that included costs, ability to upgrade and voluntary program effectiveness. Other factors were also used to show that current designated uses were inappropriate, including: Factor 2 (natural, ephemeral, intermittent, or low-flow conditions or water levels prevent the attainment of the use); Factor 4 (dams, diversions, or other types of hydrologic modifications preclude the attainment of the use); and Factor 5 (physical conditions related to the natural features of the waterbody preclude attainment of aquatic life protection uses). Although the Bay Program looked at the costs and social impact of not meeting the criteria for designated uses, it was not a driving factor (F-1). The primary reason for this was that there was not a clear analytical framework for evaluation of costs and/or social impacts in urban or agricultural settings, two important land uses in the watershed (F-2).

The Chesapeake Bay Program used a science-based, collective decision-making process to decide what data and information was necessary to support the UAA (S&T-1, S&T-3, L-4, and L-5). The required information evolved throughout the process with input from stakeholders and regulators as no UAA had previously been conducted at such a scale (S&T-1). Extensive monitoring had been conducted in the Bay since the inception of the Chesapeake Bay Program in 1983. Therefore, physical, chemical and biological data for the water, living resources and the air were available bay wide over decadal scales. After the refined Bay-specific aquatic life criteria for DO, water clarity and chlorophyll *a* were drafted, monitoring data were also used to assess whether or not the refined criteria were attainable.

Four different models were used to support the UAA (S&T-4). An airshed model was used to generate nutrient loads associated with atmospheric deposition. The Chesapeake Bay watershed model calculated nitrogen and phosphorus loads under three scenarios: all-forest; pristine; and limit-of-technology pollution controls. The model showed that agriculture was a major source of nutrients and sediments to the Bay and urban runoff also provided sediments to the Bay. The Chesapeake Bay estuary hydrodynamic and water quality models were used to assess whether or not the refined designated uses and associated numeric criteria could be met within the Bay and its tidal tributaries under these different loading scenarios. The models also revealed that even with all the technologically feasible controls (assuming unlimited funding), the previous state water quality standards for DO were still not attainable.

The UAA technical study also examined paleoecological records, which demonstrated that certain natural conditions, such as climate-driven variability in freshwater flow to the Bay, resulted in impairments of the previous state water quality standards. Thus, the paleoecological examination was proof of 131.10(g) Factor 1 (naturally occurring pollutant concentrations prevent the attainment of the use) for this UAA, and supported the argument for Factor 2 (natural, ephemeral, intermittent, or low-flow conditions or water levels prevent the attainment of the use).

No economic analysis guidance was used for the UAA because the available information did not adequately address the urban and agricultural sectors, which were significant in this effort (F-1, F-2). The Bay Program partners, therefore, developed a process to conduct economic analyses for three levels of nutrient and sediment control beyond what had been spent through the year 2000 (S&T-5). The Bay Program partners developed "implied costs per household" assuming that all costs would be paid by households in the watershed (which is unrealistic). It was noted that the perception of affordability depends on funding (e.g., with unlimited federal funding cost would not be an issue, but without federal funding, upgrades would be very costly to the communities within the watershed).

The Bay Program worked throughout the process to ensure that the concerns of the effected communities, environmental groups, wastewater treatment plant operators, and local elected officials were addressed (PA-1, PA-2). Community members were informed through numerous meetings with stakeholders, open Bay Program meetings, an extensive and comprehensive website, numerous articles published in the Alliance for Chesapeake Bay's monthly Bay Journal (watershed-wide circulation of 50,000+) and other newspapers, and presentations and publications by Bay researchers and universities in a variety of venues (PA-4). Advocacy groups were also involved in the process who understood that states could not calculate TMDLs using the original standards. The technical support document and economic analysis went through peer review and revisions, and then public review. All comments were cataloged and answered through the appropriate committee. This extensive outreach may be why there was very little political or public opposition to the UAA (PA-1, PA-4).

The documentation for the Chesapeake Bay UAA effort was extensive (L-2) and included:

- A technical support document that provided information on the refined uses and proposed criteria (U.S. EPA, 2003a); This document was amended with information on the cap needed for nutrient loads to the bay to attain the refined uses (U.S. EPA, 2004b);
- An economic analysis that was conducted to evaluate the assumed financial impact on households in the watershed (U.S. EPA, 2003b);
- A biological evaluation that was conducted to ensure that the criteria would protect endangered and sensitive species (U.S. EPA, 2003d);
- A Biological Opinion that was authored by NOAA Fisheries to confirm that the refined standards would not adversely affect endangered species (NOAA 2004);
- U.S. EPA publication of new regional criteria for DO, clarity, and chlorophyll *a* (U.S. EPA, 2004a, 2003c); and
- Extensive information on the Chesapeake Bay Program's website.

5.3.3 Resolution

The large-scale effort brought about a regulatory process for changes in the states surrounding Chesapeake Bay. The finding that natural, human-caused and nonremediable sources of pollution caused the nonattainment of water quality standards in the Bay led to the identification of five designated uses for aquatic life protection for the Bay:

- Migratory fish spawning and nursery designated use;
- Shallow-water bay grass designated use;

- Open-water fish and shellfish designated use;
- Deep-water seasonal fish and shellfish designated use; and
- Deep-channel seasonal refuge designated use.

Different DO, chlorophyll *a* and water clarity criteria were derived to specifically support these individual designated uses and were given temporal application. For example, open-water fish and shellfish use applies all year round, whereas migratory fish spawning and nursery use specific criteria apply from February 1 through May 31 (R-5).

The Chesapeake Bay UAA was conducted with the intention of adopting consistent, attainable standards across the four jurisdictions sharing the Bay's tidal waters, providing a common, scientifically based definition of restored Bay water quality (S&T-1, S&T-5, PA-1). This was successfully accomplished.

The cost of this 3-year UAA effort was nearly one million dollars, not considering the multi-million dollar monitoring and modeling effort that had preceded and supported this UAA (F-4). The key factors leading to the success of this UAA were the extensive and early involvement of and outreach to stakeholders, agencies and communities throughout the watershed (S&T-1, L-5, R-3, R-4, and PA-1). The application of the watershed and hydrodynamic/water quality models, and the use of a unique technology (paleoecological record review), supported definition of natural conditions and the determination that current uses were not attainable (S&T-4).

5.4 Cuyahoga River Ship Channel

The Cuyahoga River is located in Northeast Ohio and empties into Lake Erie. Throughout most of the last century the Cuyahoga River (Figure 5-4) has been plagued with high-profile pollution, having caught fire several times before the inception of the Clean Water Act in 1972. In the lower reach of the Cuyahoga River is the commercialCuyahoga Ship Channel, which plays an important role in the economy of Cleveland, Ohio. Because of its pollution problems, U.S. EPA classified the lower reaches of the Cuyahoga River as one of 43 Great Lakes Areas of Concern.

The Cuyahoga River Ship Channel's history of human impact has left it extremely low in DO levels. Without forfeiting its use as a navigable ship channel, the Cuyahoga River Ship Channel is incapable of supporting a warmwater habitat aquatic life use designation year round. Ohio EPA, together with the Cuyahoga Remedial Action Plan (RAP) Coordinating Committee, conducted a UAA to appropriately assign an aquatic life use to the channel (S&T-1). The end result was a site-specific partial use designation and corresponding water quality criterion that recognized both the existing use of the channel for commercial shipping and its seasonal use by migratory fish.

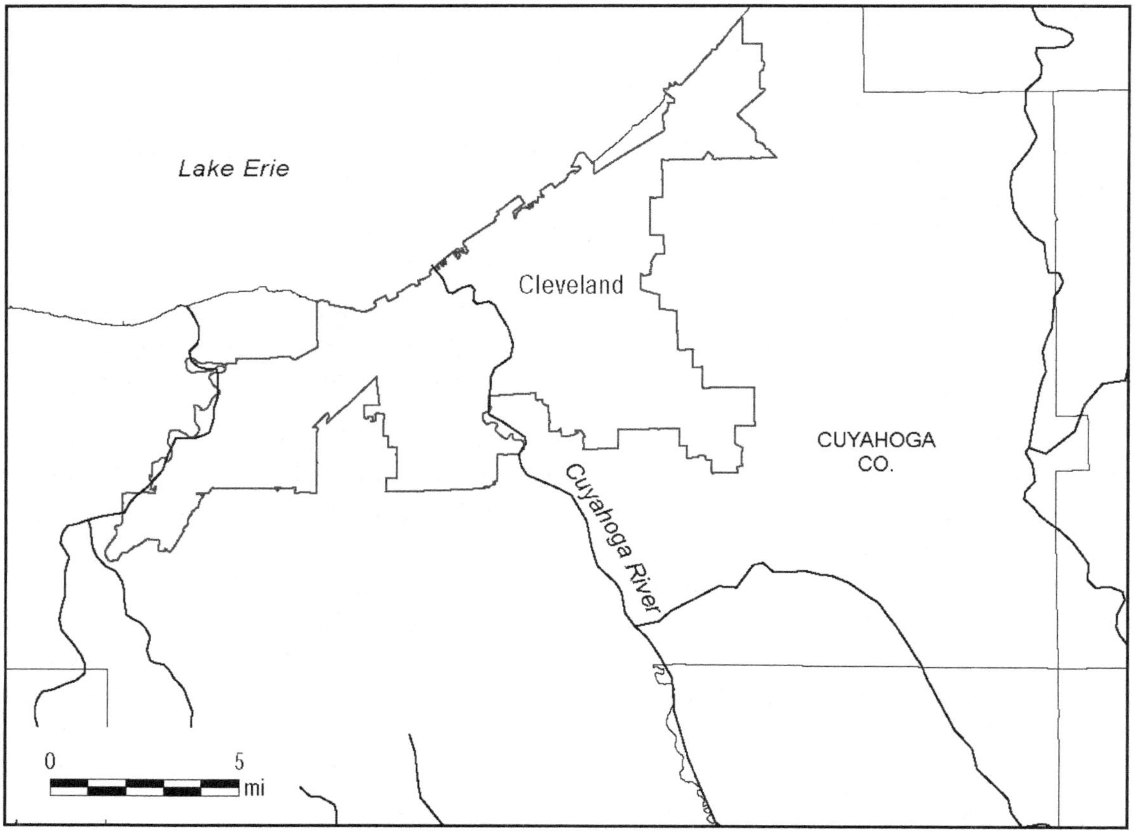

Figure 5-4. Cuyahoga River Study Area.

5.4.1 Background

During the 1970s, Ohio EPA temporarily designated some of its most polluted waterways as limited warmwater habitat use. The limited warmwater habitat use has less stringent criteria than the warmwater habitat use assigned to healthier waters. It was the intent of Ohio EPA to reassign these waters to a more appropriate use (e.g., warmwater habitat) after federal grant monies were made available to better treat the sources of pollution (i.e., wastewater treatment plants). When the grant money came in, most waterbodies (but not the Cuyahoga River) were upgraded to higher uses through the UAA process. In fact, this reassessment of waterbodies designated with the limited warmwater habitat use was the impetus for what has evolved to be a very efficient and well-defined rule making process for UAAs in Ohio. Ohio's streamlined process uses biocriteria to classify tiered aquatic life uses (R-1). Further discussion on Ohio's approach is provided in Chapter 6.0.

Unfortunately, the Cuyahoga River Ship Channel was overlooked and was never reassessed for its appropriate aquatic life use. Finally, after strong encouragement from U.S. EPA, Ohio EPA moved forward with a UAA to determine the channel's appropriate use.

5.4.2 Conducting the UAA

Ohio EPA developed the Cuyahoga RAP Coordinating Committee to oversee the remedial activities of the Cuyahoga River, including the Cuyahoga Ship Channel UAA process (L-5). The Cuyahoga RAP Coordinating Committee is made up of a 33-member task force including local, state, and federal agencies, business and industry representatives, and community interest groups. The RAP process was developed as part of the Great Lakes Water

Quality Agreement (1985) between Canada and the United States to restore the designated uses of the Areas of Concern.

Together, the Cuyahoga RAP Coordinating Committee and Ohio EPA studied the Cuyahoga River Ship Channel: historical records were assessed; the Army Corps of Engineers conducted fish surveys; and hydraulic studies, benthic surveys, fish electroshocking, and field surveys were conducted. From these studies, it was clear that the Channel habitat was stressed because of low DO levels (i.e., DO occasionally reached 1 mg/l and lower). The studies also discovered that during the spring months when flows were higher, the channel is used by fish as a migratory route. Therefore, careful consideration was needed to protect this aquatic life resource.

Ohio EPA led the effort to determine what it would take to get the channel to meet the Warmwater Habitat use (24-h average DO = 5 mg/l; minimum DO = 4 mg/l) by extensively modeling the Cuyahoga River Ship Channel using the Water Quality Analysis Simulation Program (WASP4; version 4; Ambrose et al., 1988) model (S&T-4). The results of the modeling effort were that the 23-foot deep, slow-moving channel (the retention time for the 5.6-mile course is about 10 days) would need to be decreased to a river depth of twelve feet to achieve the DO criteria. However, at this shallow depth the channel would not be able to be used for commercial shipping. The modeling results also showed that it would not be possible to restore the ship channel to conditions similar to other Lake Erie River mouths because of considerable human-induced alteration that already existed.

In addition to the modeling, a simple cost-benefit analysis was developed to understand the consequences of eliminating the channel. The results of the cost-benefit analysis made it clear that the two major steel companies that employed thousands of locals and other smaller businesses would be devastated if the ship channel were to be eliminated (S&T-5).

Because of the obvious impracticability of removing deep-water navigation from the channel, Ohio EPA proposed a new use based on Factor 3 (human caused conditions or sources of pollution prevent the attainment of the use) (F-1). Public outreach efforts and the involvement of the Cuyahoga RAP Coordinating Committee from the beginning and throughout the entire process, together with the partnership Ohio EPA had forged with the committee, led to a relatively smooth and noncontroversial UAA process.

5.4.3 Resolution

The finding that "irretrievable human induced conditions" [Ohio Administrative Code (OAC) 3745-1-26] precluded the attainment of the warmwater habitat use, together with the fact that the channel is a migratory fish passage in the spring, required that a special use designation for the Cuyahoga Ship Channel be developed that addressed the existence of both of these conditions. The final aquatic life use designation for the Cuyahoga River Ship Channel is as follows:

- During the months of June through January, when river flow is low, the use shall be limited resource water – navigation maintenance; and
- During the months of February through May, when the river flow is high, the use shall be fish passage. Fish passage is defined as "rivers and or other waterbodies that have been the subject of use attainability analysis and have been found to be incapable of supporting and maintaining a balanced, integrated, adaptive community

of water organisms but are capable of supporting the passage of warmwater fish during migratory periods."

A new criterion also had to be developed that supported the new use. From the studies and the modeling, it was found that the DO level that supported the existing condition of the ship channel was a minimum DO of 1.5 mg/l during June through January, and during the remaining months of the year whenever the river flow is less than 703 cubic feet per second. During the months of February through May whenever the river flow equals or exceeds 703 cubic feet per second, the criteria are the same as the warmwater habitat criteria (24-h average DO = 5 mg/l; minimum DO = 4 mg/l), with the exception that the biological criteria do not apply.

While establishing the new use and criteria for the Cuyahoga River Ship Channel, it was fully recognized that the DO criteria would not always be met. Consequently, the Cuyahoga RAP Coordinating Committee was held responsible for utilizing the TMDL approach to progress towards attainment of the DO criteria (S&T-6). As recognized by Ohio rules (OAC 3745-1-26), the TMDL approach must be used to enhance the DO of the ship channel "through means other than additional point and nonpoint source load reductions." Therefore, the Cuyahoga RAP Coordinating Committee is now working on alternatives such as implementing off channel re-aeration, sediment remediation, and flow augmentation to raise DO levels in the ship channel.

The Cuyahoga River Ship Channel flows through the heart of Cleveland, Ohio. Many people have a special interest in the fate of the channel, yet each person's interest is not the same. Without the support and coordination of the Cuyahoga RAP Coordinating Committee, which included 33 members representing stakeholders from business and industry, watershed and community groups, and regulatory agencies, this process could have been dead before it even started (PA-1). With everyone at the table from the beginning, the interests of all parties have been addressed in a conciliatory process (PA-4).

5.5 Spokane River

The Spokane River UAA (Figure 5-5) was initiated by a consortium of nine municipal and industrial dischargers. These parties were facing a Washington Department of Ecology ("Ecology") TMDL process that was heading in a direction that would require the dischargers to remove all their discharges from the river during the June through October time period (S&T-6, PA-6). Preliminary estimates of $700M to $1B for all point sources to comply with this requirement was a major driver for the UAA (PA-6), but the sponsors also believed that the water quality standards that the TMDL was trying to achieve were not appropriate or attainable. This is a useful case study because it reinforces most of the findings and recommendations of this WERF research project.

5.5.1 Background

The UAA was initiated in early 2003 by nine sponsors, consisting of local industrial and municipal dischargers to the Spokane River from the Lake Coeur d'Alene outlet in Idaho to Long Lake Reservoir Dam in Washington (see Figure 5-5). The need for the dischargers to "get out of the river" from Ecology's perspective was primarily driven by the aquatic life designated uses and associated D) criteria, one of which was that the cumulative effect of dischargers cannot cause the DO concentration to decrease by more than 0.2 mg/l in lakes and reservoirs, including in the lower layer in a stratified reservoir like Long Lake Reservoir (F-5, PA-6). Because Ecology's model predicted that the cumulative effect of the dischargers would violate these criteria, the draft TMDL concluded that there was no assimilative capacity for point sources (PA-6).

One interesting and challenging aspect of this UAA is that the UAA study was written at the same time that Ecology converted from a class-based to a use-based system of standards. In the conversion process, however, Ecology did not evaluate on a site-specific basis whether the list of uses for a given classification was appropriate for any given waterbody. Instead Ecology simply transferred all uses in the class [e.g., all Class A ("lake" class) uses were transferred to Long Lake Reservoir, which included salmonid spawning and core rearing]. This water quality standards rule-making also included, for the first time in Washington, provisions for UAAs and site-specific criteria. In addition, Ecology started developing UAA guidance while the UAA study was underway. One further complication was that U.S. EPA Region 10 ultimately disapproved portions of Ecology's water quality standards rule-making for reasons entirely separate from this UAA.

Yet another challenge, and benefit, was that the owner of the 5 hydroelectric dams in the study reach (Avista Corporation) was in process of relicensing the project with the Federal Energy Regulatory Commission (FERC). This was a challenge because the FERC process could result in changes to facilities and/or operations that could in turn affect water quality in the UAA study reach and downstream. As a result, the UAA had to attempt to predict, before the completion of the FERC process, how water quality and use attainability might be affected by any required changes. On the other hand, this was beneficial because Avista was conducting a number of extensive scientific studies, including modeling that had direct relevance to the TMDL and UAA (S&T-4).

5.5.2 Conducting the UAA

At the onset of the UAA process, an advisory committee was established that included regulatory and resource agencies, the Spokane tribe, dischargers, Avista, the Chamber of Commerce, industry, environmental groups (Sierra Club and Lands Council), Parks Commission, and the Long Lake property owners group (PA-1, PA-2, PA-4). This committee was informed via a series of public meetings and public education workshops, and a UAA-dedicated website. The UAA scope of work was shared with this committee at the outset and it was informed of technical findings as the UAA proceeded. In addition, UAA sponsors held one-on-one meetings with a number of the organizations involved (PA-4). Public input was also sought via the opportunity for stakeholders to comment on at least two iterations of publicly available draft UAA reports.

Fueled originally by the extremely high cost of compliance with the draft TMDL, and by the accelerated schedule for completion of the TMDL process, a UAA was considered by the sponsors to be the only alternative available to counter the draft TMDL (PA-6). The City of Spokane, being the single largest discharger to the river, would have been significantly impacted by the draft TMDL. Although the majority of the City Council was supportive of site-specific WQS determinations; one or two members were somewhat critical because of initial perception that the intent of a UAA is to remove existing uses (PA-3). This initial perception was largely overcome when it was understood that the UAA would not remove existing uses, but would refine the specific time periods and locations where subcategories of use occurred, such as trout spawning (PA-5). Despite the extensive public involvement component of this UAA process,

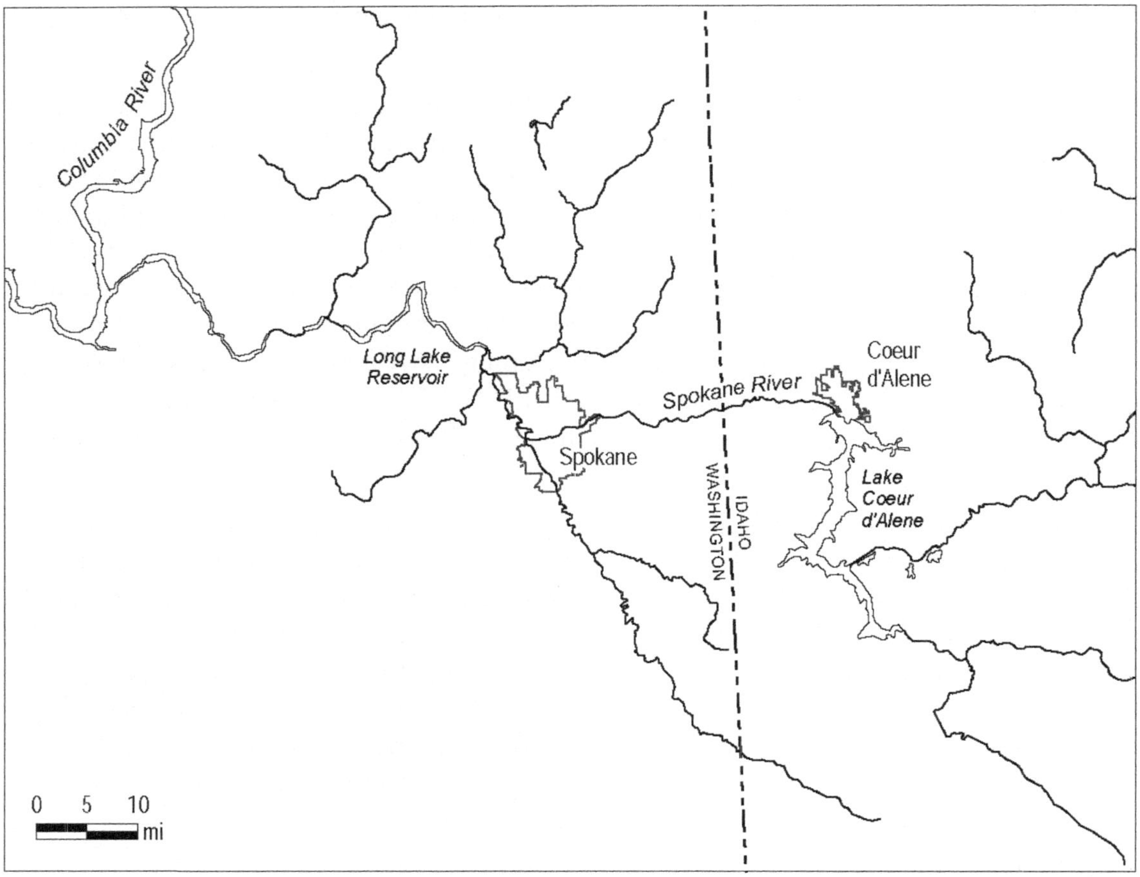

Figure 5-5. Spokane River Study Area.

environmental groups (mainly the local Sierra Club chapter) remained opposed to doing a UAA; they portrayed it to the press and to selected politicians as a process that removes existing uses so that dischargers can continue to "pollute" the river (PA-3).

Regarding agency involvement, the UAA sponsors understood that ultimately the Director of the Department of Ecology makes the final decision at the State level (S&T-3, L-1). However, the sponsors worked with all levels of Ecology, including regional staff and management, headquarters staff in the WQS unit, and upper-level water quality program managers (R-2). Although Ecology was not supportive of adopting the final recommendations in the UAA report, this ongoing dialogue assisted in the ultimate resolution of the TMDL issues (L-3, L-4). In addition, the Washington Department of Fish and Game was included in the UAA advisory committee and recognized that the river and reservoir are a mixed fishery (not just cold water), and, based on their written comments, accepted the technical factual information presented in the UAA (R-3).

The Spokane River UAA evaluated the biological and ecological basis for the designation of existing and attainable uses and recommends use designations and criteria appropriate to the Spokane River and Long Lake Reservoir. The free-flowing reach of the Spokane River between Post Falls Dam and the Stateline provides a variety of habitat types (pool, riffle, run) that support rainbow trout spawning from early April through late May or June. In contrast, Long Lake Reservoir is currently managed as a mixed fishery, including naturally reproducing populations of cool- and warm-water fishes as an inherited (that is, not stocked) tournament and recreational bass fishery.

Based on extensive available physical, chemical, and biological data (over 200 existing documents, over 60 years of research (L-2), extensive water quality modeling [using Ecology's CE-QUAL-W2 model for the Spokane River and Long Lake Reservoir]) (S&T-4), the UAA proposed that two new subcategories of use be established in Washington's surface water quality standards (R-5). These uses are referred to as "Spokane River Cold- and Cool-Water Mixed Fishery" and "Long Lake Reservoir Mixed Fishery." These subcategories provide for:

- the protection of indigenous and nonindigenous cold- and coolwater fish species and other associated aquatic life;
- limited spawning, rearing, and migration of rainbow trout; and
- a more site-specific temporal and geographic application of biologically based DO criteria.

The Spokane River UAA biological assessment showed that Ecology's current surface water quality criteria (9.5 mg/l absolute DO level and 0.2 mg/l differential DO level) cannot be met a high percentage of the time, regardless of source treatment scenario, including removing the point source loads entirely from the Spokane River (S&T-4). It does show that the criteria proposed in the Spokane River UAA (protective of existing and attainable uses) can be met under most conditions. The UAA also recommends a comprehensive implementation plan to control both point and nonpoint sources of pollutants to the Spokane River, including implementing technologies to achieve 95% reduction in the point source phosphorus load to the Spokane River.

Two of the 131.10(g) factors were cited in the UAA in support of the recommended use refinements: Factor 4 (dams, diversions or other types of hydrologic modifications preclude the attainment of the use); and Factor 5 (physical conditions related to the natural features of the waterbody preclude attainment of aquatic life protection uses). These two technical factors were supplemented with cost and economic assessments in the UAA, including documenting the economic impacts on lower income households and documenting marginal costs versus marginal environmental benefits (S&T-5, F-1, and F-2). The UAA sponsors decided not to cite Factor 6 (substantial and widespread economic and social impacts) because they felt that the analyses behind the technical factors alone were strong enough to support the recommended use refinements (F-1, F-2).

Consideration of downstream water quality standards was a key issue in this UAA (L-6). The Spokane River enters the reservation boundary of the Spokane Tribe a relatively short distance downstream of the Long Lake Dam. Extensive water quality modeling (S&T-4) demonstrated that the water quality improvements that the sponsors proposed in the UAA would improve downstream DO, but the Tribe's DO standard would not be met at all times even if all discharges to the Spokane River were totally removed. Thus, it appears that the downstream DO issues will likely have to be resolved via the FERC relicensing process (e.g., reservoir or tailrace aeration).

5.5.3 Resolution

The UAA study was completed in December 2004. The UAA provided the backup for a rule-making petition to refine the designated uses and criteria to those recommended in the UAA report. The rule-making petition was formally filed by the sponsors in late December 2004. In early 2005, Ecology wrote a letter to the sponsors asking that the petition be withdrawn without prejudice in exchange for a commitment to continue good faith negotiations on key elements of the TMDL. The sponsors agreed to this request in February 2005 and a Steering Committee was formed to facilitate development of a TMDL collaboration process. This TMDL collaboration process led to development a decision document entitled: "Foundational Concepts" for implementation of the TMDL in June 2006. Members of the Collaboration included Ecology, Spokane County, cities of Spokane and Spokane Valley, Liberty Lake Water and Sewer District, Kaiser Aluminum, and Inland Empire Paper; with participation by the Sierra Club and Lands Council, Avista, and the Spokane Tribe. U.S. EPA participated as an observer.

The Foundational Concepts is a detailed and comprehensive water quality improvement plan that includes an adaptive implementation mechanism. A key element of this plan is the explicit opportunity to revisit the questions of water quality standards appropriateness and attainability at the 10-year critical review and assessment milestone of the TMDL process.

5.6 Lower French Gulch and Blue River, Colorado

Mining operations near the Town of Breckenridge, Colorado, resulted in toxic concentrations of metals caused by acid mine drainage, as well as dramatic impacts on aquatic habitat in French Gulch (Figure 5-6). A comprehensive stakeholder group came together to address 303(d) listings under the Clean Water Act as well as regulatory requirements of a CERCLA remediation effort (S&T-1, S&T-6, and L-5). The impracticality of a complete cleanup led the stakeholders to consider a UAA (PA-6). The UAA combined extensive scientific and technical assessments, economic realities of a remediation effort, and a facilitated stakeholder process (S&T-5, F-1, and PA-1) (Lane Wyatt et al., 2003). The outcome of the UAA included site-specific criteria for dissolved metals and revised use designations (R-5).

5.6.1 Background

French Gulch is a mining area near the Town of Breckenridge in Summit County, Colorado. Mining for gold and other metals was conducted in the Wellington-Oro underground mine from 1882 to 1930 and again from 1940 to 1973. Mining using large floating dredge boats also took place on the surface, removing gold from fluvial gravel deposits.

The upper portion of French Gulch, upstream of the Wellington-Oro Mine, supports a healthy fishery of native Colorado River cutthroat trout. Water quality sampling showed acute toxic conditions for trout in French Gulch from the mining area to the confluence with the Blue River, approximately a 2- to 3-mile stretch. This stream segment was included on the 303(d) list as impaired because of pH, cadmium, and zinc. Zinc and cadmium concentrations were also above the aquatic life standard on the Blue River from French Gulch to Swan River. The benthic organism community diversity was extremely low and was almost entirely composed of chironomids (midges, tolerant of poor water quality).

A coordinated groundwater and surface water monitoring program indicated drainage from the underground mine was the primary contributor of metals loadings to French Gulch. Acid mine drainage is formed when sulfide-bearing minerals are exposed to and react with air and water to form sulfuric acid. The acid mine water then dissolves metals, such as cadmium and zinc, and exits the mine through seeps and enters French Gulch. The seeps contain zinc concentrations of 123,000 μg/l, typically 5000 times that found in French Gulch upstream of the mine. The seeps flow year round at an average rate of 100 gallons per minute. In spring they may flows as high as 500 gallons per minute. The above-ground placer mining had dramatic impacts.

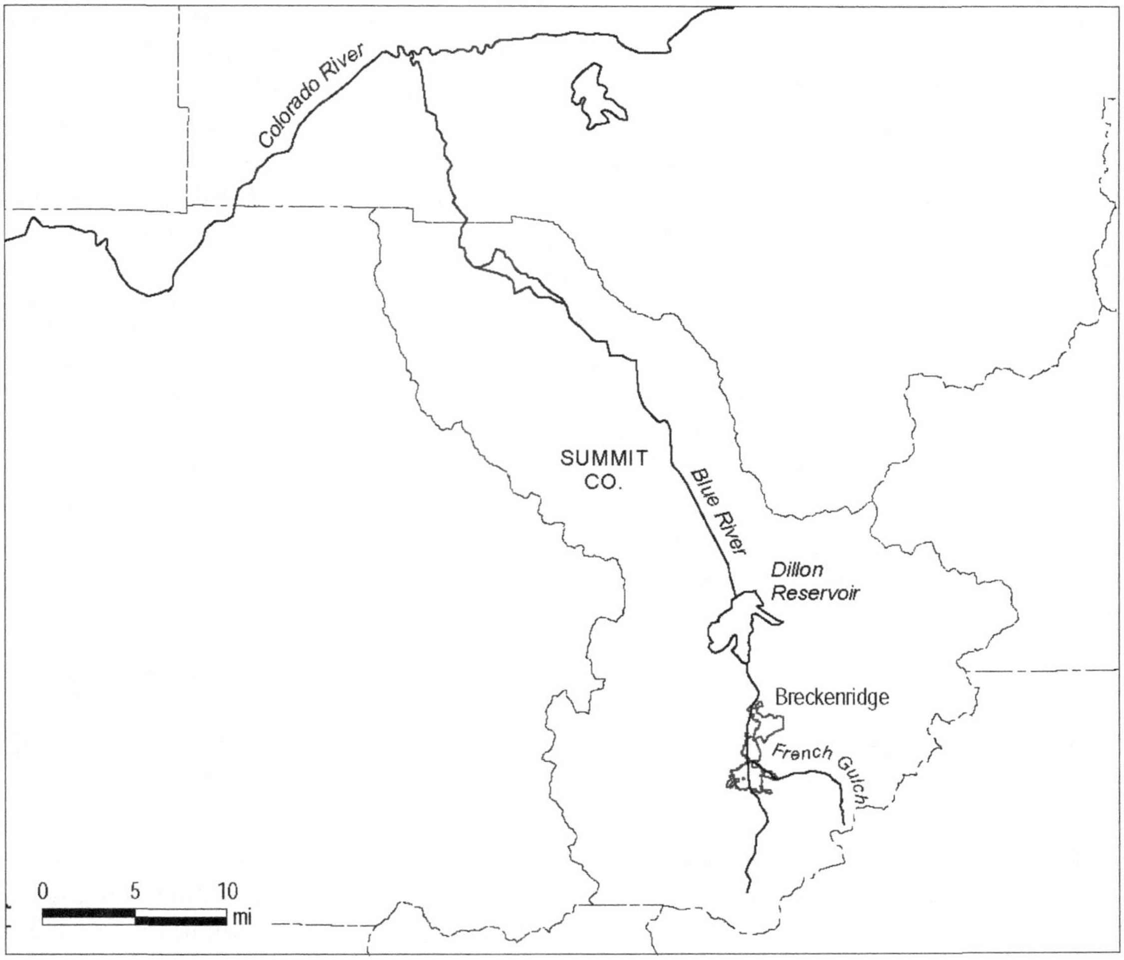

Figure 5-6. Lower French Gulch and Blue River, Colorado, Watershed.

on habitat as well, largely destroying natural habitat and leaving piles of large gravel and cobble in the stream. In some places the stream disappears as it flows through these piles

5.6.2 Conducting the UAA

Initial investigations were conducted by the U.S. EPA and the Colorado Department of Public Health and Environment (CDPHE) through the Clean Water Act Nonpoint Source program and the Rocky Mountain Headwaters program in 1989. Other participants included Colorado's Division of Minerals and Geology and the Northwest Colorado Council of Governments. The scope of the potential clean-up of the Wellington-Oro Mine and restoration of French Gulch was determined to exceed the Nonpoint Source program's available resources, so the state and U.S. EPA decided to take a multiprogram and community-based approach (S&T-1). In 1994 U.S. EPA awarded Northwest Colorado Council of Governments a grant for the development of a methodology for establishing aquatic biological goals for areas affected by acid mine drainage. A multi-agency and community-based group was formed to manage the studies and remediation (PA-1). The group was called the French Gulch Remediation Opportunities Group, or FROG. The group contracted with an independent facilitator to address issues of: resources to fund clean-up; remediation activities; historic preservation; and community involvement. Sixteen governmental agencies participated in FROG, including: U.S.

EPA; CDPHE; Colorado Division of Minerals and Geology; State of Colorado's Engineer's Office and Division of Wildlife; United States Geological Survey; United States Forest Service; Summit County Commissioners, Planning Department, Water Quality Committee, and Environmental Health Department; Town of Breckenridge manager, engineer, and planner; Northwest Colorado Council of Governments; Blue River Water Commissioner; Upper Blue Planning Commission; and the District Attorney's Office (S&T-1, L-5, and R-3). The owner of the mines, B&B Mines, and other mining, development, tourism, and historic preservation interests also participated along with area residents.

The FROG worked together to define the goal of the UAA and develop the approach (S&T-3). The goal was defined simply as developing technically justifiable site-specific water quality standards for the stream segments under consideration. The State of Colorado recognizes a UAA as one of the appropriate methods to support site-specific standards [5 CCR. Section 1002-31, Section 31.7(1)(b)(iii)] (L-1). The approach used in developing the UAA was designed to be consistent with the State's definition of a UAA [Section 31.5(30)]: "an assessment of the factors affecting the attainment of aquatic life uses or other designated uses, which may include physical, chemical, biological, and economic factors." The approach used in developing this UAA can be summarized as follows:

- Characterize existing conditions in terms of aquatic life, physical habitat, stream hydrology and water column chemistry;
- Attempt to project future water quality conditions that are expected to exist after the implementation of the CERCLA water treatment plant at the Wellington-Oro Mine site;
- Use literature information and professional judgment to determine the aquatic life potential for the stream segments in the study area recognizing future water quality conditions and any existing physical, biologic or hydrologic limitations to aquatic life; and
- Propose site-specific water quality standards that are protective of potential aquatic life.

The UAA provided a detailed characterization of water quality and a comparison to relevant water quality standards for several monitoring locations, where data were available for the study area. Information from this description of chemical and hydrologic conditions was used to depict future stream chemistry based on anticipated metals removal capabilities of the CERCLA remedial project.

The UAA used available data and information to characterize the existing aquatic communities (both trout populations and macroinvertebrate communities) and physical habitat conditions in both French Gulch and the Blue River, including seasonal stream flows, aquatic habitat, and channel benthic substrates. A detailed literature review on the effects of dissolved species of zinc, lead, and cadmium on aquatic biota, including trout and macroinvertebrates, was performed to assess the aquatic life potential of the stream segments under consideration. Fish and wildlife agencies were extensively involved in this process (R-3). Information on existing habitat and literature information on aquatic community potential was used to determine the potential aquatic communities that could be expected to exist if water-quality conditions improved as a result of the Wellington-Oro Mine remedial treatment facility. Conclusions from this analysis then were used to propose site-specific water-quality standards for the stream segments under consideration by the UAA.

The economic considerations of the UAA were limited to existing relevant information on the costs of both proposed and potential water quality and physical habitat improvements in the impacted stream segments. Sources included the following:

- The Engineering Evaluation/Cost Analysis for the CERCLA evaluation of the Wellington-Oro Mine site;
- Construction bid proposals for the restoration of selected stream reaches of the Blue River near the confluence with the Swan River;
- The Prospective Purchaser Agreement that was an integral part of the proposed remediation of the Wellington-Oro Mine; and
- An understanding of habitat limitations.

Taken in combination, this information provided an understanding of the financial costs of implementing the proposed recommendation as well as first-order estimates of financial limitations associated with improved aquatic habitat in French Gulch (F-2).

5.6.3 Resolution

In May 2003, the Summit Water Quality Committee submitted the UAA to the Colorado Water Quality Control Commission. The UAA concluded that aquatic habitat was severely impacted by historic dredge mining. The revisions to the water quality standards were approved by the Colorado Water Quality Control Commission on September 8, 2003, and are summarized below.

In French Gulch, below the Wellington-Oro Mine to the confluence with the Blue River, criteria for cadmium, zinc, and lead were revised and set as "existing quality." Existing quality, as defined in the state regulations, means the 85th percentile of the available data for dissolved metals. The designated use for this segment (segment 11 of the Blue River basin) was also revised from "reviewable" to "use protected" within the context of the state's anti-degradation rules. A stream designated as use protected is not subject to antidegradation review (R-5).

In the Blue River, the stream segment between the confluence with French Gulch and the confluence with Swan River, previously segment 2, was divided into two segments, 2a and 2b. Habitat modifications in the upper portion of the segment, 2a, caused by mining and construction of a kayak course within the Town of Breckenridge, were much more significant than downstream. The habitat modifications in the upper portion result in an absence of spawning and rearing habitat for aquatic species. Below the Town fishery habitat improves markedly in segment 2b and additional habitat improvement projects are under consideration. Because of the habitat differences, different life stages were expected to be present in each segment (L-6). Consequently, different toxicity based cadmium and zinc criteria were promulgated. For segment 2a, the cadmium criterion was set to 4 ug/l and the zinc criterion was hardness dependent and calculated. For segment 2b, both cadmium and zinc criterion were hardness dependent and calculated. The criteria were based on zinc and cadmium toxicity to the different life stages of brown trout that were expected to occur in segments 2a and 2b. The designated use for segment 2a was also revised to "use protected" (R-5).

These criteria formed the basis for treatment targets for the remediation effort. The selected treatment alternative for the Wellington-Oro involves chemical precipitation accomplished via lime addition. Significant water quality improvement is anticipated in French Gulch itself with the initiation of treatment plant operations. However, the Commission determined that additional water quality improvement beyond that accomplished through collection and treatment of mine water at the Wellington-Oro site is infeasible as a result of irreversible anthropogenic causes.

The resolution of the water quality standards to attainable levels led to continued cooperative efforts to improve water quality and restore the mine site. In 2004 voters in Breckenridge and Summit County approved the purchase of the 1,800-acre mine property. The $9 million Golden Horseshoe Open Space purchase addressed the scope of cleanup actions, including construction of the water treatment plant at the Wellington-Oro Mine. In addition to this cooperative land acquisition, Summit County and Breckenridge have worked together to plan and develop the mine property and restore the Blue River downstream from French Gulch.

CHAPTER 6.0

STATEWIDE UAA PROTOCOLS

This chapter describes five statewide UAA protocols that were included in the research of the 54 UAA efforts nationwide. In addition, an example of a state with a tiered aquatic life use system, and three examples of states that have developed legislation to address designated uses during wet weather are presented.

In general, the UAA protocols described in this chapter were developed in response to statewide use designation issues. The protocols serve to streamline UAA efforts, in some instances where numerous UAAs had to be conducted in a short time period, and may save costs.

Several of the research findings presented in Chapter 3.0 relate directly to the use of statewide UAA protocols:

- Finding F-5: When state UAA protocols are available, the process becomes streamlined and the costs for conducting the UAA are typically lower.
- Finding R-1: When statewide UAA protocols and/or guidance exist, state agencies tended to be well-prepared and the UAA process proceeds more smoothly.
- Finding PA-6: The risk of spending large amounts of money in the face of tight TMDL schedules created an urgency (or willingness) to move forward with UAA analyses.

The statewide UAA protocols discussed in this chapter address both recreational and aquatic life uses. In addition, legislation to address designated uses during wet weather is summarized for three states. The examples are presented as follows and are summarized in Table 6-1:

- Recreational UAA Protocols
  - Colorado
  - Kansas
  - Missouri
- Aquatic Life UAAs
  - Ohio
  - Oklahoma
  - Maine
- State Legislation to Address Wet-Weather Uses
  - Indiana
  - Maine
  - Massachusetts

Table 6-1. Overview of Statewide UAA Protocols and State Legislation to Address Wet-Weather Issues Discussed in Chapter 6.0.

Type	State	Overview
Recreational Use UAA Protocol	Colorado	Recreational UAA protocol developed in 2003 in response to landowner concerns related to public access; 50 to 100 UAAs completed and approved to date.
	Kansas	U.S. EPA rejected recreational use changes proposed by the state in 1998; Agreement required state to conduct UAAs for more than 1,000 waterbodies; recreational use UAA protocol developed in 2001; Majority of UAAs have been approved by U.S. EPA.
	Missouri	Water quality standards revisions would apply whole body contact recreational use to all classified waters of the state; change would have required 911 dischargers to provide disinfection at an estimated cost of $300 million; State developed a recreational UAA protocol to apply to the 500 streams receiving discharges; First round of UAAs completed in 2005 and approved. U.S. EPA review resulted in disapproval of a number of use removals leading to changes in protocol and reevaluation of some streams.
Aquatic Life Use UAA Protocol	Ohio	Designated uses are reviewed for each waterbody during the state's rotating basin monitoring and assessment program; Uses are assessed using a biosurvey program and narrative biocriteria; State has a system of tiered aquatic life uses; Rotating monitoring and assessment program schedule coincides with NPDES permitting schedule and TMDL development schedule.
	Oklahoma	Protocols were developed ~20 years ago for aquatic life use UAAs; "One-day UAAs" rely on gathering of desktop information and one-day field surveys to assess the Fish and Wildlife Propagation subcategories of Warmwater Aquatic Life Community and Habitat Limited Aquatic Community; Hundreds of UAAs have been conducted since the protocols were developed.
Tiered Aquatic Life Use System	Maine	A tiered aquatic life use system based on biological condition has been in place since before the Clean Water Act; assessment of the state's biocriteria associated with the uses is based on benthic macroinvertebrate monitoring.
State Legislation to Address Wet-Weather Issues	Indiana	Draft rule issued in May 2006 that implements Senate Act 620 (SEA) passed in 2005; SEA 620 allows for establishment of a CSO wet-weather limited use subcategory of the recreational designated use that applies during and for up to four days following the end of the CSO discharge.
	Maine	State employs a variance approach to address CSO conditions in wet weather; Temporary CSO subcategory can be established that requires CSO community to submit a UAA demonstrating that the use cannot be attained; water quality standards may be temporarily suspended or modified for CSO events beyond a rainfall-selected event size.
	Massachusetts	State allows for both variances in water quality standards and use reclassification for CSO impacted waters; a partial use CSO subcategory was established in 1996 for receiving waters with a substantial number of outfalls; application of this use requires approval of a UAA.

6.1 Recreational UAA Protocols

Protocols for conducting UAAs for recreational uses from Colorado, Kansas, and Missouri were included in the 54 UAA efforts researched and are described below.

6.1.1 Colorado

In Colorado UAA guidance was developed for recreational use UAAs in response to private landowner concerns statewide. In 2000 statements made by the Colorado Water Quality Control Commission related to public accessibility for streams designated as primary contact recreation raised concerns by landowners over property rights. These concerns led to development of the UAA guidance.

The Recreational Use Classification Guidance (CDPHE, 2003) was developed in a cooperative effort between a group of stakeholders and the Colorado Department of Public Health and the Environment. The process for evaluating recreational uses outlined in the guidance consists of three steps: (1) identifying existing uses; (2) assessing potential for water play by children; and (3) assessing potential for other primary contact recreation, considering access (physical and proximity to developed areas) and physical conditions (e.g., water depth, flow conditions, frequency of flow, and water temperature). A two-page worksheet is provided with the guidance to assist with collection of information necessary for the UAA.

The recreational UAAs following this guidance are being led by private landowners in an attempt to limit public accessibility to streams on their property by removing the primary contact recreation use. Most of the UAAs are being conducted on small and intermittent streams in rural areas. The key factor influencing use attainability has been low-flow conditions in most instances.

CDPHE provides guidance during development of the UAAs. Final UAA proposals are reviewed by CDPHE, which makes a recommendation to the State's Water Quality Control Commission. The Commission then approves or rejects the recommendation, and sends the UAA to U.S. EPA for final review.

To date, between 50 and 100 recreational use UAAs have been conducted using the State guidance. All have been approved by U.S. EPA. Municipalities are now showing more interest in recreational UAAs. However, the State has so far not pursued or approved any UAAs in an urban setting.

6.1.2 Kansas

Kansas Department of Health and Environment (KDHE) originally proposed secondary contact recreational use changes that were rejected by the U.S. EPA in 1998. U.S. EPA objected to this approach arguing that, at a minimum, all Clean Water Act Section 101(a) uses, including primary contact recreation, should automatically be assigned to any stream, lake or wetland lacking a formal UAA. In March 2001 KDHE and U.S. EPA negotiated a memorandum of understanding and a schedule to conduct UAAs on those streams and lakes that were previously assigned secondary contact recreation use. In spring 2001 KDHE began to conduct recreational UAAs on surface water segments designated for secondary contact recreation for which assessments had not previously been completed (1,456 surface water segments).

On April 13, 2001, Governor Bill Graves signed into law Senate Bill 204 (K.S.A. 2001 Supp. 82a-2001), which became effective on September 1, 2001. Essentially, this state law changed the definition of classification and use designation of streams only. For a stream to be classified it needed to meet the following statutory requirements: 1) streams with a median flow of equal to or greater than 1 cfs; or 2) streams receiving a wastewater discharge from a NPDES permitted facility; or 3) streams actually inhabited by threatened or endangered aquatic species; or 4) streams shall be classified if studies conducted on streams less than 1 cfs provide important refuges for aquatic life. Also, the law required KDHE to complete recreational UAAs on all stream segments of the state where a recreational UAA had not been completed before October 31, 2005.

During the 2003 session of the Kansas Legislature, House Bill 2219 was passed. This bill, effective May 1, 2003, amended K.S.A. 2001 Supp. 82a-2001 (SB 204). The statute established three subclasses for primary contact recreational use and two subclasses for secondary contact recreational use. The three subclasses under which a stream will be designated primary contact recreation (swimming, kayaking, etc.) depend on the level of recreation expected. Primary class-A is assigned to those streams designated public swimming areas. Primary class-B is assigned to those streams where moderate full body contact recreation is expected and the stream is open to and accessible to the public. Primary class-C is assigned to those streams where moderate full body contact recreation is infrequent and the stream is not open to and accessible to the public under Kansas law. Secondary class-A is assigned to streams where the water depth is not sufficient enough to support primary contact recreation and is open to and accessible to the public. Secondary class-B is assigned to streams where the water depth is not sufficient enough to support primary contact recreation and it is not open to and accessible to the public under Kansas law.

For the greater than 2000 stream recreational UAAs conducted by KDHE, Factor 2 (natural flow conditions) was typically the deciding factor. Average water depth of at least 0.5 m and maximum water depth of 1.0 m during base flow are considered necessary to support primary contact recreation (KDHE, 2001). U.S. EPA has approved the majority of UAAs submitted by KDHE. There has been some disagreement between U.S. EPA and KDHE over UAAs that resulted in the assignment of secondary contact recreational use. In these instances, U.S. EPA believes either the stream depth was sufficient to support primary contact recreation or the stream was accessible by the public for recreational activities like swimming. KDHE is completing additional field surveys to investigate these streams (Thompson, 2007).

6.1.3 Missouri

Missouri Department of Natural Resources (MDNR) proposed changes to the State's bacteria standards that would result in the whole body contact recreation (WBCR) use being assigned to all classified waters of the State. This new use classification would require 911 discharging facilities to provide disinfection of their effluent at an estimated cost of $300 million statewide. Anticipating the need to conduct numerous UAAs to prevent certain waterbodies from receiving the default WBCR, MDNR developed a protocol in 2004 to apply to streams to determine if they meet or do not meet minimum depth requirements for WBCR.

The Recreational Use Attainability Analysis Protocol (MDNR, 2004) was applied to approximately 500 of the state's classified waters where there were known facilities discharging bacteria to intermittent streams. Nearly all of the UAAs were based on an assessment of how low-flow conditions and insufficient depth prevent attainment of the WBCR use. The UAAs were handled through the State's rulemaking process with the rules becoming final in September 2005.

The protocol addressed the need for streamlining the UAA process and providing a well-defined approach for UAAs conducted by both state staff and other stakeholders. The protocol was developed with input from U.S. EPA and was published on September 29, 2004, and approved on November 3, 2004. The protocol consisted of four pages to be completed and relied on ocular depth information at three points (encouraged to be easily accessible) within every 5 miles of a stream segment.

MDNR had a very clear understanding of the costs associated with conducting so many UAAs. They developed a work plan based on the process that KDHE went through (see discussion in previous section) and estimated the total number of hours needed to complete the UAAs. Staff worked overtime to complete the UAAs. The review process (staff report; public comment; internal review committee; review by the Clean Water Commission) was also understood and planned for. MDNR estimated that each UAA would, on average, take 2.5 hours to complete (1.5 hours of travel plus 20 minutes for each of three stops for the waterbody segment).

The UAA process included submittal of the UAA by MDNR staff or stakeholder, followed by public comment, review of the UAA by an internal review committee of three people in MDNR, and then review by the Clean Water Commission, which at times included additional public comment. MDNR staff or their contractors conducted approximately 90% of the UAAs and played the primary role in reviewing the UAAs, including obtaining public comment; municipalities and industries conducted about 10% of the UAAs. U.S. EPA Region 7 provided comments on the UAAs that influenced staff's recommendations. Missouri's Stream Teams, a working partnership of citizens concerned about Missouri streams, also played a role by being informed of the process. The Clean Water Commission played a significant role in determining the final outcome for UAAs where there was significant public comment.

There was considerable political interest in the UAA process, especially when the staff reported that disinfection could cost Missouri facilities $300 million. The agricultural communities and property rights groups were particularly concerned that the rulemaking could result in cattle being fenced out of the streams. The state held numerous public meetings that were attended by State legislators. This contributed to the rulemaking process proceeding smoothly.

In addition to the public meetings, MDNR maintained a user-friendly website that described the rulemaking and UAA. The community was also informed as part of the stakeholder committees that MDNR established for drafting the rules and the protocol, as well as attending Clean Water Commission meetings. MDNR also provided briefings to Missouri's Stream Teams. Results of the UAAs were also published in newspapers, etc. for public comment. The public had opportunity to comment on the rulemakings that established WBCR unless a UAA was done, development of the protocol, the UAAs, and the internal review committee's findings.

Overall, the process that the State of Missouri followed resulted in 141 waterbody segments where WBCR was not designated based on insufficient depth. In its October 2006 review of the rulemaking, U.S. EPA evaluated the UAAs and the supporting data provided by MDNR to determine whether the UAAs were sufficient to make a technically and legally defensible demonstration that the WBCR use is not attainable. U.S. EPA determined that adequate documentation existed to support this determination for only 42 of the 141 waterbody segments. U.S. EPA determined that a WBCR use designation was appropriate for the other 99 waterbody segments. In conducting this evaluation, U.S. EPA noted that the data were not adequate to provide the necessary scientific and technical rationale in the administrative record supporting MDNR's decision. For these waters, U.S. EPA identified four types of issues associated with the data: 1) data were collected during drought conditions; 2) locational information indicated that the data contained in the UAA were not from the correct classified stream segment; 3) data were insufficient; and 4) instances where data were inconclusive.

In response to U.S. EPA's disapproval of 99 UAAs, MDNR began a process of revisiting their UAA protocol in the beginning of 2007 to make revisions to address concerns raised by U.S. EPA and received during public comment. These revisions are intended to address the representativeness of sites selected for the UAA, interviews and documentation of an existing use; and significance of data collected or observed in the field. A key change in the protocol is that actual depth measurements will be required for the UAA rather than relying on visual observations. MDNR plans to revisit the UAAs for all 99 waterbody segments disapproved by U.S. EPA and collect additional data during the 2007 recreational season.

6.2 Aquatic Life UAAs

Protocols for conducting UAAs for aquatic life uses from Ohio and Oklahoma were included in the 54 UAA efforts researched and are described below. In addition, Maine's system of tiered aquatic life uses is described.

6.2.1 Ohio

The State of Ohio has developed a system that has evolved since the late 1970s to inventory streams and creeks. Originally, best professional judgment was used to designate uses on a particular waterbody. Since then, an assessment process that includes a tiered aquatic life use assessment and biocriteria index has been developed. Because an impairment assessment cannot be made until the "use" question is answered, the state assesses the designated uses of the water with these systematic assessment methods before allowing a permitted discharge on a waterbody. The state is doing an inventory of all creeks throughout Ohio with a U.S. Geological Survey hydrologic unit code of 11 digits or less.

Ohio's aquatic life use designations were originally adopted in 1978 and are based on the ecological potential of the waterbody. The state's current tiered aquatic life uses are as follows:

- Warmwater Habitat: basic restoration goals for most streams;
- Limited Warmwater Habitat: temporarily designated in 1978 water quality standards as not meeting warmwater, being phased out;
- Exceptional Warmwater Habitat: capable of supporting and maintaining and exceptional or unusual community of warmwater organisms;
- Modified Warmwater Habitat: attainable condition for streams under drainage maintenance or other essentially permanent hydromodification;
- Seasonal Salmonid Habitat: supports lake run steelhead trout fisheries;
- Coldwater Habitat: supports native cool- and coldwater species and put-and-take trout stocking; and
- Limited Resource Waters: essentially irretrievable, human induced (e.g., widespread watershed modification) or naturally occurring conditions (e.g., ephemeral flow).

To assign a waterbody one of the tiered uses, Ohio established a biosurvey system based on narrative bioassessment criteria. Essentially, field surveys are conducted to determine the potential use of the waterbody using a biological criteria index. The narrative criteria evolved into numeric biocriteria, which were adopted into Ohio Water Quality Standards in 1990.

When conducting UAAs, many agencies have found it difficult to identify the necessary steps and what is needed to support one of the six U.S. EPA conditions for approving a UAA. Ohio has found many benefits to linking continuous monitoring with use assessment. Having a structured statewide monitoring, assessment and tiered aquatic life use system allows for these questions to be easily answered in a streamlined process. In the case for Ohio, their monitoring, assessment, and tiered aquatic life use system has led to the following benefits:

- identifies the benchmarks for tiered aquatic life uses;
- outlines the method for defining criteria and the reference condition;
- recognizes different expectations for different classes of habitats;
- distinguishes "stressors" from human activities;
- considers global and regional extinction of species; and
- considers biological, chemical, and physical conditions.

Since 1978, Ohio has conducted over 1,000 aquatic life use designation revisions.

6.2.2 Oklahoma

Oklahoma has developed an efficient method of assessing waters and processing UAAs, which are often referred to as "One-day UAAs." The process begins with gathering desk-top information (e.g., USGS maps) about the waterbody and then conducting a one-day field survey following the specific protocols outlined in the "Unified Protocols for Beneficial Use Assignment for Oklahoma Wadable Streams" (Oklahoma Water Resources Board, 2001). Oklahoma's One-day UAAs have evolved over the last 15 years since the survey has gone from a "subjective, sensory inspection (appearances and smells) to its current state of quantitative assessment that involves physical, chemical, and biological aspects of the stream."

The protocols outline how the Fish and Wildlife Propagation subcategories of Warmwater Aquatic Life Community and Habitat Limited Aquatic Community are assigned. Other use designations would require a more time-intensive UAA process. Additionally, the protocols outline the specific physical, chemical, and biological measurements that should be taken, the laboratory analysis that should be done, and a description of the findings that may be discovered.

Approximately 30 UAAs have been conducted in the past 10 years, and hundreds have been conducted in the 20 years since the protocols were developed.

6.2.3 Maine

Although the State of Maine does not have a statewide protocol for developing UAAs for aquatic life uses, it does have a system of tiered aquatic life uses based on a gradient of biological condition. A tiered use system with accompanying water quality criteria specific to each use is useful when UAAs are conducted, as it provides options for revising uses as appropriate for a specific waterbody within a state's regulatory framework.

U.S. EPA (2005) has stated that a system of tiered uses could:

- Accommodate observable differences in expected biological condition in waterbodies in different ecological regions; and

- ♦ Provide an objective means of describing the biological potential for a specific waterbody.

Maine's tiered aquatic life use system has been in place and in practice in Maine's water quality laws before the adoption of the Clean Water Act. Although their system has gone through an evolutionary process, Maine now has four tiers of aquatic life uses that represent the acceptable biological conditions that may occur in the state. These different water quality classes are presented in Table 6-2.

Maine has a strong monitoring program which supports their assessment system. Waters within one of Maine's five river basins are sampled on a rotating basis every 5 years. A standardized biologically based monitoring protocol has been followed since 1983. In 2003, the Maine Board of Environmental Protection adopted numeric freshwater biocriteria into rule. These criteria are based on assessment of benthic macroinvertebrate data collected in the monitoring program.

Other states with a system of tiered aquatic life uses based on a gradient of biological integrity include Ohio (see Section 6.2.1) and Washington. Washington's tiered system was finalized and incorporated into the state's water quality standards on December 31, 2006.

Table 6-2. Maine's Water Quality Classification.

CLASS	MANAGEMENT	BIOLOGICAL STANDARD
AA*	High quality water for recreation and ecological interests. No discharges or impoundments permitted.	Habitat shall be characterized as natural and free flowing. Aquatic life shall be as naturally occurs.
A	High quality water with limited human interference. Discharges limited to noncontact process water or highly treated wastewater of quality equal to or better than the receiving water. Impoundments allowed.	Habitat shall be characterized as natural. Aquatic life shall be as naturally occurs.
B	Good quality water. Discharge of well-treated effluent with ample dilution permitted. Impoundments allowed.	Habitat shall be characterized as unimpaired. Discharges shall not cause adverse impacts to aquatic life. Receiving water shall be of sufficient quality to support all aquatic species indigenous to the receiving water without detrimental changes in the resident biological community.
C	Acceptable water quality. Maintains the interim goals of the federal Water Quality Act (fishable/swimmable). Discharge of well-treated effluent permitted impoundments allowed.	Habitat for fish and other aquatic life. Discharges may cause some changes to aquatic life, provided that the receiving waters shall be of sufficient quality to support all species of fish indigenous to the receiving water and maintain the structure and function of the resident biological community.
Impoundments	Riverine impoundments not classified as Great Ponds and managed for hydropower generation.	Support all species of fish indigenous to those waters and maintain the structure and function of the resident biological community.

Note: *The narrative aquatic life standard is the same for Class AA and Class A.
Source: U.S. EPA, 2005.

6.3 Legislation to Address Uses During Wet Weather

The examples presented in this section are of statewide legislation that has been promulgated to specifically address designated uses during wet weather. Controlling combined sewer overflows (CSO), stormwater, and nonpoint source pollution to attain recreational water quality standards during wet weather is often unrealistic and/or unaffordable. The challenge of wet-weather water quality standards attainment is a prominent topic nationwide. It is currently being addressed directly and indirectly in many state programs either by direct modification to the state standard or by implicit interpretation of existing standards. The examples provided below for the States of Indiana, Maine, and Massachusetts directly provide for exceptions to standards to address CSO conditions using the following methods:

- Indiana – temporary use suspension with a limited wet-weather use subcategory;
- Maine – variance approach with a temporary use subcategory; and
- Massachusetts – variance and use reclassification with partial or revised uses.

6.3.1 Indiana

Indiana has been aggressive in trying to create CSO exceptions for water quality standards through legislative changes in the standard. These exceptions are necessary because Indiana's water quality standards require that *E. coli* concentrations not exceed 235 counts per 100 mL at any time, making it difficult to prepare a CSO LTCP that would comply with water quality standards. In 2000, the Indiana General Assembly enacted SEA 431, which provided a format for coordinating CSO LTCPs with any necessary water quality standards reviews, and established a process for temporarily suspending full body contact recreation after the LTCP is implemented.

Deficiencies in SEA 431 were identified and clarified by interested parties in 2005. In that year, SEA 620 was sponsored and became law (NACWA, 2005). Key elements of SEA 620 follow below:

- Establishes a CSO wet-weather limited use subcategory of the recreational designated use that applies during and for a period of up to four days following the end of the CSO discharge. The water quality-based requirements associated with the subcategory will be determined by the approved LTCP;
- Allows for an integrated review of a LTCP and UAA, if a change in use designation is needed;
- Provides that CSO controls must be cost-effective and affordable;
- Establishes effective State-based enforceable mechanisms to protect communities during the time before LTCPs are fully implemented and sets forth a process by which the permit can serve as the enforceable mechanism. The primary tool is a compliance schedule for meeting water quality-based requirements associated with CSOs during LTCP development, approval and implementation; and
- Revises the variance statute so that the duration of a variance is tied to the permit term and can be renewed to cover the entire LTCP implementation period.

The Indiana Department of Environmental Management (IDEM) issued a draft rule in May 2006 that implements SEA 620. IDEM is incorporating changes to address various comments, including those from U.S. EPA. IDEM intends to issue this as a final rule for U.S. EPA's approval.

In September 2006, the City of Indianapolis submitted its final CSO LTCP to IDEM and U.S. EPA for review and approval. The LTCP included a recreational UAA that requested a wet-weather limited use subcategory of the recreational use consistent with SEA 620 (City of Indianapolis Department of Public Works, 2006). In October 2006, a consent decree was filed in U.S. District court that outlined the city's agreement with IDEM and U.S. EPA following their review of the LTCP and asked that the court accept the settlement. The settlement was accepted by the court in December 2006. The LTCP was approved by IDEM in January 2007, except for the UAA, which remains under review by the agency. Regarding the UAA, the consent decree outlines a timeframe for IDEM to begin the water quality standards review process or issue a final decision that such a revision will not occur (*United States and State of Indiana v. City of Indianapolis*, Civ. No. 1:06-cv-1456).

6.3.2 Maine

The State of Maine employs a variance approach to address CSO conditions in wet weather. Maine passed legislation codifying standard procedures for providing water quality standards variances to CSO dischargers during implementation of an approved LTCP (MDEP, 2003). A temporary CSO subcategory is established and a citizen board, appointed by the Governor, approves the creation of the subcategory. The CSO community is required to submit a LTCP, an implementation schedule, and a UAA demonstrating that the use cannot be attained. The Board may then temporarily suspend or modify the water quality standards associated with the use, including the extent and duration, for CSO events beyond a selected rainfall event size.

The state had been assisting two communities, Augusta and Bangor, with developing permits based on this approach using grant funding. However, political issues arose in these towns related to differences in the timeframe for compliance of the communities using the wet-weather standards versus those required for neighboring industries that made Augusta and Bangor decide to stop pursuing this approach. The State also decided that they would no longer assist communities with the approach in the future; however, it is still available for other permittees to pursue (True, 2007).

6.3.3 Massachusetts

The State of Massachusetts allows for both a variance in water quality standards and use reclassification for CSO-impacted waters (MA DEP, 2000). A variance for a CSO is used when additional data need to be collected to understand sources of bacteria, impacts on water quality, and effectiveness of controls or while progress is being made in CSO controls. Variances allow for short-term modifications of Massachusetts' water quality standards when interim control measures or further analyses are warranted. Thus, variances allow communities to comply with temporary water quality standards in their NPDES permits while progress is being made to comply with the existing standards. Variances are issued by the MA DEP and can be both discharger- and pollutant-specific, and are time-limited; they do not change the current waterbody class designation (e.g., Class B).

Massachusetts amended its water quality standards in 1996 to include a CSO subclassification in its use classification system for receiving waters with substantial numbers of CSO outfalls. The application of a CSO subclassification requires U.S. EPA approval of a UAA. Massachusetts uses the UAA to evaluate the attainability of the designated use, particularly whether CSO controls would likely cause substantial and widespread economic and social impact. MA DEP grants partial use designations, based on results from UAA, in CSO-impacted waters where MA DEP is certain that the designated uses or standards cannot and will not be attained on a permanent basis. Partial use generally indicates a transient impairment of uses and can be defined by seasons or a particular storm event when a use such as primary public recreation contact and bathing will be unattainable in CSO-impacted waters. The use must be fully protected downstream, in other seasons, or during smaller storm events. In areas where MA DEP determines that designated uses cannot and will not be met on a permanent basis, MA DEP will then consider a change in classification from Class B to Class C. This option must be based on UAA findings that the Class B designated use cannot be reasonably attained.

The Class B (CSO) Massachusetts subclassification requires that CSO controls be sufficient to meet water quality standards 95% of the time, generally no more than four CSOs per year. A UAA must be developed that demonstrates achieving greater than 95% control would cause substantial and widespread economic and social impact. Lesser controls expected to only achieve criteria 75–95% of the time require legislative approval. The Commonwealth must make the UAA available for public review and comment and receive U.S. EPA approval before applying the Class B (CSO) subclassification to a particular waterbody. The standard suspends only the bacteriological criteria; toxic pollutants are not affected.

MA DEP has conducted several UAAs, with the most notable being for Boston Harbor in 1997 (see case study in Chapter 5.0). However, two obstacles to reclassification have been noted:

- If there are sensitive areas, reclassification to a CSO category is not allowed. This has been the primary issue associated with reclassifying the Charles River. U.S. EPA Region 1 contends that the Charles River should be used for swimming, which is part of the reason why the MWRA pursued a long-term variance for the Charles River as contrasted with a permanent reclassification. U.S. EPA and MA DEP have given MWRA regulatory assurance through a stipulation filed in federal court that MWRA will be allowed a series of 3-year variances for 15 years. During this period, MWRA will not be required to implement additional CSO control measures for discharges to the Charles River unless new information is made available indicating that water quality conditions in the river are significantly different than what is currently known and predicted (Kubiak, 2007).
- MA DEP has indicated that the U.S. EPA regulatory hurdles associated with the CSO categories have been significant. U.S. EPA Region 1 has indicated that this designation should only be made when it is the only financially feasible choice. As a result, communities may be pushed to eliminate CSOs for all or a portion of the 5% of discharge time still remaining if U.S. EPA determines that the community can afford it (Brander, 2004).

CHAPTER 7.0

IDENTIFIED NEEDS

The research contained in this report provides new insights into factors leading to success for use attainability studies. However, as the research has shown, UAAs are still far from routine and pose technical, regulatory, political, and process challenges. Outlined in this chapter are four areas of research and development that could further advance the effectiveness of the UAA process: 1) a comprehensive database of UAA case studies; 2) improved guidance manuals on the UAA process; 3) better technical guidance on water quality standards; and 4) dissemination of findings and tools.

7.1 UAA Database

Success is best built off learning from others' successes and failures. To facilitate that learning, a comprehensive national database of UAA case studies is needed that can be searched by specific issues, locations, or topics. Although this research compiled and evaluated many UAA case studies at considerable effort and time, relevant information is often limited and not readily available to the profession or public. A comprehensive national UAA database could be CD- or web-based. It should include a comprehensive characterization of the issues and process used for a particular UAA, a simple description of the UAA, and links to full documentation. Using this database, interested parties could identify their own specific issues and then search the database to identify other UAA studies with similar challenges.

Also needed are more detailed case studies of successful UAAs, readily available to interested parties. This research provided several case studies, but more covering a wider range of locations, parameters, and UAA topics are needed. Case studies should include not only what was done, but also the specific elements of the UAA that contributed to its success. These case studies should be indexed to the database to facilitate easier identification.

The need for accessible documentation on more case studies was highlighted in a November 6, 2006 conference call among WERF subscribers interested in this issue, as well as in a December 6-7, 2006, WERF subscriber workshop in Baltimore exploring research challenges. This need was also quite evident in the conduct of this research as the research team found it challenging to identify suitable case studies for this project.

7.2 Guidance Manuals

The UAA process involves a diverse range of considerations, each with its own technical and procedural challenges. Additional guidance in the following topical areas would be valuable to provide interested parties with practical knowledge derived from case studies:

- state standardized protocols;
- public involvement processes and strategies;
- economic and affordability analyses;

- defining "feasible";
- evaluating physical constraints to uses;
- wet-weather evaluations;
- defining "existing uses";
- defining "natural conditions";
- evaluating downstream uses;
- weight of evidence approaches; and
- categorical UAAs.

The research contained in this report points to these areas as important contributors to the success or failure of UAAs performed to date. The UAA process would benefit from more detailed guidance on how to approach these topics based on case study experiences, critical review of U.S. EPA guidance and regulations and, where needed, new research and assessments. This need was emphasized in the December 2006 WERF subscriber meeting.

7.3 Technical Guidance on Standards

The focus of the UAA process is, of course, on determining water quality standards. Although water quality standards have been developed and utilized over the past 35 years, we have many complicating issues that are not universally well understood even today. Examples of these issues include

- scientifically defensible bacterial standards based on risk;
- wet-weather considerations;
- high- and low-flow conditions;
- refined or tiered uses; and
- assessment of compliance with criteria (location, frequency, magnitude, and duration).

Each of these areas would benefit from technical guidance that outlines the issues, regulatory constraints, alternative approaches, and experiences from different states. In addition, the guidance should also include evaluations of new approaches.

7.4 Dissemination of Findings and Tools

One final need identified is the wide dissemination of these research findings and any future information on UAA case studies, tools, assessments and guidance. This need was underlined in both the November 2006 and December 2006 WERF subscriber discussions. There is a need to make more information available to WERF subscribers, but also to regulators, nongovernmental organizations, and the public. Suggestions to meet this need have included documentation on case studies, an information clearinghouse, presentations in a variety of venues to reach out to all stakeholders, and journal articles. Educational outreach is as much a key to improving the UAA process as is the need for more research.

References

Ambrose, R.B. et al. W*ASP4, A Hydrodynamic and Water Quality Model. Model Theory, User's Manual, and Programmer's Guide*; EPA/600/3-87-039. US EPA: Athens, GA, 1988.

Brander, K. CSO Coordinator, Massachusetts Department of Environmental Protection. Personal communication, 2004.

CH2M HILL. Spokane River and Long Lake Reservoir Use Attainability Analysis. CH2M HILL: Boise, ID, 2004.

Chesapeake Bay Program. *Chesapeake 2000 Agreement.* http://www.chesapeakebay.net/agreement.htm.

Chesapeake Bay Program. *Watershed Profile, 2006.* http://www.chesapeakebay.net/wspv31/(guipzi45tzxrcdqgdlil4455)/WspAbout.aspx?basno=1&topic=5.

City of Indianapolis Department of Public Works. Raw Sewage Overflow Long Term Control Plan and Water Quality Improvement Report. City of Indianapolis Department of Public Works: Indianapolis, IN, 2006.

Colorado Department of Public Health and Environment, Water Quality Control Division (CDPHE). *Recreational Use Classification Guidance*; version 1.1. CDPHE: Denver, CO, 2003.

Kansas Department of Health and Environment, Bureau of Environmental Field Services. (KDHE). *Guidance Document for Use Attainability Analyses (UAAs)*. KDHE: Topeka, KS, 2001.

King, E. S. Memo from EPA Office of Science and Technology Director to Regional Water Division Directors; Improving the Effectiveness of the Use Attainability Analysis (UAA) Process; March 13, 2006. http://www.epa.gov/waterscience/standards/uaa/pdf/memo_king.pdf.

Kruzen, A. *Ozark Sierran* 2005, *37(3)* [Online]. http://missouri.sierraclub.org/SierranOnline/2005/07/uuas.htm.

Kubiak, D. Massachusetts Water Reclamation Authority. Personal communication, January 19, 2007.

Lane Wyatt, Summit Water Quality Committee, TDS Consulting Inc., and Walsh Aquatic Consultants, Inc. Use-Attainability Analysis, Lower French Gulch and the Blue River Downstream From French Gulch Near Breckenridge, Summit County, Colorado; Prepared on Behalf of Summit Water Quality Committee, U.S. Environmental Protection Agency, and CDPHE. Lane Wyatt, Summit Water Quality Committee: Silverthorne, Colo., 2003.

Los Angeles Regional Water Quality Control Board. Proposed Amendment to the Water Quality Control Plan – Los Angeles Region to Suspend the Recreational Beneficial Uses in Engineered Channels during Unsafe Wet Weather Conditions. Los Angeles RWQCB: Los Angeles, CA, 2003. http://63.199.216.5/bpa/bpa_detail.php?rbResNo=2003-010&no=17.

Maine Department of Environmental Protection. *Maine Rev. Statutes Annotated,* 2003, *38,* Sect. 464.

Massachusetts Department of Environmental Protection. *Guidance for Abatement of Pollution from CSO Discharges*. MDEP, 1997. http://www.mass.gov/dep/water/laws/csoguide.pdf.

Massachusetts Department of Environmental Protection. *Code Mass. Regul.,*2000, *314,* 4.00.

Missouri Department of Natural Resources. Water Protection Protocol Recreational Use Attainability Analysis Protocol. MDNR: Jefferson City, MO, 2004.

National Academy of Sciences. *Assessing the TMDL Approach to Water Quality Management.* National Academy Press: Washington, DC, 2001.

National Oceanic and Atmospheric Agency. National Marine Fisheries Service Endangered Species Act Biological Opinion: Ambient Water Quality Criteria for Dissolved Oxygen, Water Clarity and Chlorophyll a for the Chesapeake Bay and Its Tidal Tributaries; Northeast Region; F/NER/2003/00961. NOAA: Silver Spring, MD, April 19, 2004.

Oklahoma Water Resources Board. *Unified Protocols for Beneficial Use Assignment for Oklahoma Wadeable Streams.* Oklahoma Water Resources Board: Oklahoma City, OK, 2001.

Thompson, C. Bureau of Environmental Field Services, Kansas Department of Health and Environment. Personal communication, January 19, 2007.

True, J.M. CSO Coordinator, Division of Water Quality Management, Maine Department of Environmental Protection. Personal communication, January 19, 2007.

United States and State of Indiana v. City of Indianapolis, Civ. No. 1:06-cv-1456.

U.S. Environmental Protection Agency (U.S. EPA). *Combined Sewer Overflow Control Policy;* EPA 830-B-94-001. *Fed. Regist.* 1994, *59,* 18688, 18693.

U.S. EPA. Interim Economic Guidance for Water Quality Standards: Workbook; EPA-823-B-95-002. U.S. EPA: Washington, DC, 1995.

U.S. EPA. *Fed. Regist*. 1998, *63*, 36,756.

U.S. EPA. Technical Support Document for Identification of Chesapeake Bay Designated Uses and Attainability; EPA 903-R-03-004. Region III, Chesapeake Bay Program Office, U.S. EPA: Washington, DC, 2003a.

U.S. EPA. Economic Analysis of Nutrient and Sediment Reduction Actions to Restore Chesapeake Bay Water Quality. Region III, Chesapeake Bay Program Office, U.S. EPA: Washington, DC, 2003b.

U.S. EPA. Ambient Water Quality Criteria for Dissolved Oxygen, Water Clarity and Chlorophyll a for the Chesapeake Bay and Its Tidal Tributaries; EPA 903-R-03-002. Region III, Chesapeake Bay Program Office and Water Protection Division, U.S. EPA: Washington, DC, 2003c.

U.S. EPA. Biological Evaluation for the Issuance of Ambient Water Quality Criteria for Dissolved Oxygen, Water Clarity and Chlorophyll a for the Chesapeake Bay and Its Tidal Tributaries. Region III, U.S. EPA: Washington, DC, 2003d.

U.S. EPA. Ambient Water Quality Criteria for Dissolved Oxygen, Water Clarity and Chlorophyll a for the Chesapeake Bay and Its Tidal Tributaries: 2004 Addendum. Region III, Chesapeake Bay Program Office and Water Protection Division, U.S. EPA: Washington, DC, 2004a.

U.S. EPA. Technical Support Document for Identification of Chesapeake Bay Designated Uses and Attainability: 2004 Addendum. Region III, Chesapeake Bay Program Office and Water Protection Division; U.S. EPA: Washington, DC, 2004b.

U.S. EPA. *Use of Biological Information to Better Define Designated Aquatic Life Uses in State and Tribal Water Quality Standards: Tiered Aquatic Life Uses,* Draft. U.S. EPA: Washington, DC, 2005.

Water Environment Research Foundation and the National Association of Clean Water Agencies. *Collaborative Water Quality Solutions: Exploring Use Attainability Analyses*; WERF; NACWA: Alexandria, VA, 2005.

APPENDIX A

RESOURCE PANEL INFORMATION

Panel Members

Arid West Water Quality Research Program

Maine Department of Environmental Protection

City of Vacaville, CA

Oklahoma Water Resources Board

City of Spokane, WA

Massachusetts Department of Environmental Protection

Spokane County, WA

Florida Department of Environmental Protection

City of Independence, MO

America's Clean Water Foundation

Boise City, ID

Indiana Department of Environmental Management

Iowa Department of Natural Resources

EPA Region 4

Kansas Dept. of Health and Enviroment

New York Department of Environmental Conservation

Nevada Division of Environmental Protection

FTN Associates

APPENDIX B

CASE STUDY INVENTORY

Table B-1. Inventory of UAA Case Study - Name, State(s), and EPA Region(s).

UAA	State(s)	EPA Region(s)
Statewide Recreational UAAs, Missouri	MO	7
Wheeling UAA Demonstration Project, Ohio River, Ohio and West Virginia	OH; WV	3, 5
Aquatic Life Use Upgrades for Buck, Valley, Flint, and Lost Creeks, Alabama	AL	4
Chesapeake Bay UAA, Multiple States	DC; DE; MD; NY; PA; VA; WV	2, 3
UAAs in Lake Nutrient TMDLs, Minnesota	MN	5
Class 7 (Limited Resource Value Waters) Reclassifications, Minnesota	MN	5
City of Indianapolis UAA, Indiana	IN	5
Pittsburgh Regional UAA, Pennsylvania	PA	3
Sulphur Creek UAA, Sandoval County, New Mexico	NM	6
Christina River UAA, Delaware	DE	3
US Forest Service UAA for New World Mining District, Montana	MT	8
Denver International Airport UAA for South Platte River, Colorado	CO	8
UAA for Harrisonburg WTP Chlorine Variance, Virginia	VA	3
Little Isaacs Creek UAA, Virginia	VA	3
Lower French Gulch and Blue River UAA, Colorado	CO	8
Numerous Recreational UAAs, Colorado	CO	8
Chicago Area Waterways System UAA, Illinois	IL	5
Dell Arkansas Power Plant UAA for Drainage Ditches, Arkansas	AR	6
Marina Reclassifications for Shellfish Harvesting, North Carolina	NC	4
Starkist UAA for Pago Pago Harbor, American Samoa	AS	9
Federal Paperboard UAA for Cape Fear River, North Carolina	NC	4
Lower Boise River Tributaries UAA, Idaho	ID	10
Bucktail Creek UAA, Idaho	ID	10
City of Quincy UAA for Wasteways, Washington	WA	10
Santa Ana River UAA, California	CA	9
Marathon Oil Company UAA for Coal Draw, Wyoming	WY	8
Class 4C Isolated Effluent Statewide UAA, Wyoming	WY	8
Class 4A Artificially Constructed Canals and Ditches Statewide UAA, Wyoming	WY	8
Blackbird Creek UAA, Idaho	ID	10
Spokane River and Long Lake Reservoir UAA, Idaho and Washington	ID; WA	10
Donahue Industries UAA for Angelina River, Texas	TX	8
Malheur River UAA, Oregon	OR	10
Engineered Flood Channels UAA for Recreational Uses in Wet Weather, California	CA	9
Statewide Recreational UAAs, Kansas	KS	7
Snake River Hells Canyon UAA, Idaho and Oregon	ID, OR	10
Boston Harbor UAA for CSO Discharges, Massachusetts	MA	1

UAA	State(s)	EPA Region(s)
Statewide Aquatic Life UAAs, Ohio	OH	5
Statewide Aquatic Life UAAs, Oklahoma	OK	6
Cabin Creek UAA, Georgia	GA	4
Squaw Creek UAA, California	CA	9
Fenholloway River UAA, Florida	FL	4
New Iberia UAA, Louisiana	LA	6
Owen's Lake UAA, California	CA	9
City of Vacaville UAA for Old Alamo Creek, California	CA	9
City of San Luis Obispo UAA for San Luis Obispo Creek, California	CA	9
Naturally Impaired Waters of the Lahontan Region Use Removal, California	CA	9
East LA Lancaster POTW UAA, California	CA	9
Central Valley Selenium and Boron UAA, California	CA	9
City of Vacaville UAA for New Alamo Creek, Cache Slough, and Ulatis Creek, California	CA	9
Hydropower-Related UAAs for Ragged Lake and Seboomook Lake, Maine	ME	1
Central Valley Basin-Wide pH and Turbidity UAA, California	CA	9
Ecoregion Refinement of DO Criteria, Mermentau River Basin Example, Louisiana	LA	6
UAA for Ramos Forested Wetlands at Amelia, Louisiana	LA	6
Cuyahoga River Ship Channel UAA, Ohio	OH	5

Statewide Recreational UAAs, Missouri

State(s)	MO	**EPA Region(s)** 7	
Leading effort	Missouri Department of Natural Resources		
Contacts	Phil Schroeder, Missouri Department of Natural Resources		
Start	2004	**End**	On-going
Setting	☒ Effluent Dominated	☒ Wet-weather Impacted	☒ Urban
	☒ Other: Various settings across state, primarily low-flow streams		
Pollutant Source Category	☒ Agricultural Runoff	☒ Natural Background	☒ Urban Stormwater
	☒ CSOs	☒ POTWs	☐ Mining
	☒ Industry	☒ Other: Various sources across state, primarily POTWs	
UAA Factor -131.10 (g)	2, 3, 4, 5	☐ Undetermined	
Designated Use Assessed	☐ Aquatic Life	☒ Recreation	☐ Drinking Water
	☐ Other		
Parameters of Concern	☒ Bacteria	☐ Metals	☐ Nutrients
	☐ Organic Hydrocarbons	☐ Pesticides/Herbicides	☐ Solids
	☐ Temperature	☐ Ammonia	☐ BOD
	☐ Chloride	☐ Dissolved Oxygen	
	☐ Other		
Status	☒ Completed & Adopted	☐Considered not developed	☒On-going;
	☒ Completed & Rejected	☐ Started - abandoned	☒Under Consideration
	☒ Under EPA review		
Outcome	☐ Use Upgrade	☒ Use Removal/Revision	☒ No Change in Use
	☐Other:		

Wheeling UAA Demonstration Project, Ohio River, Ohio and West Virginia

State(s)	OH; WV	**EPA Region(s)** 3 and 5	
Leading effort	Ohio River Valley Water Sanitation Commission		
Contacts	Jason Heath, Ohio River Valley Water Sanitation Commission		
Start	August 2005	**End**	To be determined
Setting	☐ Effluent Dominated	☒ Wet-weather Impacted	☒ Urban
	☒ Other: Urban and agricultural settings		
Pollutant Source Category	☒ Agricultural Runoff	☒ Natural Background	☒ Urban Stormwater
	☒ CSOs	☐ POTWs	☒ Mining
	☐ Industry	☐ Other:	
UAA Factor -131.10 (g)	1, 3, 6	☐ Undetermined	
Designated Use Assessed	☐ Aquatic Life	☒ Recreation	☐ Drinking Water
	☐ Other		
Parameters of Concern	☒ Bacteria	☐ Metals	☐ Nutrients
	☐ Organic Hydrocarbons	☐ Pesticides/Herbicides	☐ Solids
	☐ Temperature	☐ Ammonia	☐ BOD
	☐ Chloride	☐ Dissolved Oxygen	
	☐ Other		
Status	☐ Completed & Adopted	☐ Considered not developed	☒ On-going;
	☐ Completed & Rejected	☐ Started - abandoned	☒ Under Consideration
	☐ Under EPA review		
Outcome	☐ Use Upgrade	☐ Use Removal/Revision	☐ No Change in Use
	☒ Other: Ongoing, UAA under consideration		

Aquatic Life Use Upgrades for Buck, Valley, Flint, and Lost Creeks, Alabama

State(s)	AL	**EPA Region(s)** 4	
Leading effort	Alabama Department of Environmental Management		
Contacts	Lynn Sisk, Alabama Department of Environmental Management		
Start	1998	**End**	2001
Setting	☒ Effluent Dominated	☐ Wet-weather Impacted	☒ Urban
	☒ Other: Urban and agricultural settings		
Pollutant Source Category	☒ Agricultural Runoff	☒ Natural Background	☒ Urban Stormwater
	☐ CSOs	☒ POTWs	☐ Mining
	☐ Industry	☐ Other:	
UAA Factor -131.10 (g)	2, 3, 4, 5, 6	☐ Undetermined	
Designated Use Assessed	☒ Aquatic Life	☐ Recreation	☐ Drinking Water
	☐ Other		
Parameters of Concern	☐ Bacteria	☐ Metals	☐ Nutrients
	☐ Organic Hydrocarbons	☐ Pesticides/Herbicides	☐ Solids
	☐ Temperature	☒ Ammonia	☒ BOD
	☐ Chloride	☒ Dissolved Oxygen	
	☐ Other		
Status	☒ Completed & Adopted	☐Considered not developed	☐On-going;
	☐ Completed & Rejected	☐ Started - abandoned	☐Under Consideration
	☐ Under EPA review		
Outcome	☒ Use Upgrade	☐ Use Removal/Revision	☐ No Change in Use
	☐Other:		

Chesapeake Bay UAA, Multiple States

State(s)	DC; DE; MD; NY; PA; VA; WV	**EPA Region(s)** 2 and 3	
Leading effort	EPA Region 3, Chesapeake Bay Program Office		
Contacts	Richard Batiuk, EPA Region 3		
Start	2000	**End**	2004
Setting	☐ Effluent Dominated	☐ Wet-weather Impacted	☐ Urban
	☒ Other: Large watershed, estuarine setting		
Pollutant Source Category	☒ Agricultural Runoff	☒ Natural Background	☒ Urban Stormwater
	☒ CSOs	☒ POTWs	☐ Mining
	☒ Industry	☒ Other: Various sources within watershed	
UAA Factor -131.10 (g)	1, 2, 3, 4, 5, 6	☐ Undetermined	
Designated Use Assessed	☒ Aquatic Life	☐ Recreation	☐ Drinking Water
	☐ Other		
Parameters of Concern	☐ Bacteria	☐ Metals	☒ Nutrients
	☐ Organic Hydrocarbons	☐ Pesticides/Herbicides	☒ Solids
	☐ Temperature	☐ Ammonia	☐ BOD
	☐ Chloride	☒ Dissolved Oxygen	
	☒ Other: Clarity, chlorophyll-a; submerged aquatic vegetation		
Status	☒ Completed & Adopted	☐ Considered not developed	☐ On-going;
	☐ Completed & Rejected	☐ Started - abandoned	☐ Under Consideration
	☐ Under EPA review		
Outcome	☒ Use Upgrade	☒ Use Removal/Revision	☐ No Change in Use
	☒ Other: Refined uses and regional criteria		

UAAs in Lake Nutrient TMDLs, Minnesota

State(s)	MN	**EPA Region(s)** 5	
Leading effort	Minnesota Pollution Control Agency		
Contacts	Frank Kohlasch, Minnesota Pollution Control Agency		
Start	Under consideration	**End**	NA
Setting	☐ Effluent Dominated	☐ Wet-weather Impacted	☒ Urban
	☒ Other: Primary concern is urban settings, but agricultural as well		
Pollutant Source Category	☒ Agricultural Runoff	☐ Natural Background	☒ Urban Stormwater
	☐ CSOs	☐ POTWs	☐ Mining
	☐ Industry	☐ Other:	
UAA Factor -131.10 (g)	1, 3, 5, 6	☐ Undetermined	
Designated Use Assessed	☐ Aquatic Life	☒ Recreation	☐ Drinking Water
	☐ Other		
Parameters of Concern	☐ Bacteria	☐ Metals	☒ Nutrients
	☐ Organic Hydrocarbons	☐ Pesticides/Herbicides	☐ Solids
	☐ Temperature	☐ Ammonia	☐ BOD
	☐ Chloride	☐ Dissolved Oxygen	
	☒ Other: clarity; chlorophyll-a		
Status	☐ Completed & Adopted	☐ Considered not developed	☐ On-going;
	☐ Completed & Rejected	☐ Started - abandoned	☒ Under Consideration
	☐ Under EPA review		
Outcome	☐ Use Upgrade	☐ Use Removal/Revision	☐ No Change in Use
	☒ Other: UAAs under consideration		

Class 7 (Limited Resource Value Waters) Reclassifications, Minnesota

State(s)	MN	**EPA Region(s)** 5	
Leading effort	Minnesota Pollution Control Agency		
Contacts	Gerald Blaha, Minnesota Pollution Control Agency		
Start	1980's	**End**	On-going
Setting	☒ Effluent Dominated	☐ Wet-weather Impacted	☐ Urban
	☐ Other:		
Pollutant Source Category	☒ Agricultural Runoff	☒ Natural Background	☐ Urban Stormwater
	☐ CSOs	☒ POTWs	☐ Mining
	☐ Industry	☐ Other:	
UAA Factor -131.10 (g)	2, 3, 4, 5	☐ Undetermined	
Designated Use Assessed	☒ Aquatic Life	☒ Recreation	☐ Drinking Water
	☐ Other		
Parameters of Concern	☐ Bacteria	☐ Metals	☒ Nutrients
	☐ Organic Hydrocarbons	☐ Pesticides/Herbicides	☒ Solids
	☐ Temperature	☐ Ammonia	☒ BOD
	☐ Chloride	☒ Dissolved Oxygen	
	☐ Other		
Status	☒ Completed & Adopted	☐Considered not developed	☒On-going;
	☐ Completed & Rejected	☐ Started - abandoned	☐Under Consideration
	☐ Under EPA review		
Outcome	☐ Use Upgrade	☒ Use Removal/Revision	☐ No Change in Use
	☐Other:		

City of Indianapolis UAA, Indiana

State(s)	IN	**EPA Region(s)** 5	
Leading effort	City of Indianapolis		
Contacts	Jodi Perras, Perras & Associates		
Start	2001	**End**	On-going
Setting	☐ Effluent Dominated	☒ Wet-weather Impacted	☒ Urban
	☐ Other:		
Pollutant Source Category	☒ Agricultural Runoff	☐ Natural Background	☒ Urban Stormwater
	☒ CSOs	☐ POTWs	☐ Mining
	☐ Industry	☒ Other: Septic tanks	
UAA Factor -131.10 (g)	2, 3, 4, 6	☐ Undetermined	
Designated Use Assessed	☐ Aquatic Life	☒ Recreation	☐ Drinking Water
	☐ Other		
Parameters of Concern	☒ Bacteria	☐ Metals	☐ Nutrients
	☐ Organic Hydrocarbons	☐ Pesticides/Herbicides	☐ Solids
	☐ Temperature	☐ Ammonia	☐ BOD
	☐ Chloride	☐ Dissolved Oxygen	
	☒ Other: High flows during large storm events		
Status	☐ Completed & Adopted	☐ Considered not developed	☐ On-going;
	☐ Completed & Rejected	☐ Started - abandoned	☒ Under Consideration
	☐ Under EPA review		
Outcome	☐ Use Upgrade	☐ Use Removal/Revision	☐ No Change in Use
	☒ Other: UAA under consideration		

Pittsburgh Regional UAA, Pennsylvania

State(s)	PA	**EPA Region(s)** 3	
Leading effort	EPA Region 3		
Contacts	Denise Hakowski, EPA Region 3		
Start	2004	**End**	NA
Setting	☐ Effluent Dominated ☐ Other:	☒ Wet-weather Impacted	☒ Urban
Pollutant Source Category	☒ Agricultural Runoff ☒ CSOs ☐ Industry	☐ Natural Background ☐ POTWs ☐ Other:	☒ Urban Stormwater ☒ Mining
UAA Factor -131.10 (g)		☒ Undetermined	
Designated Use Assessed	☐ Aquatic Life ☐ Other	☒ Recreation	☐ Drinking Water
Parameters of Concern	☒ Bacteria ☐ Organic Hydrocarbons ☐ Temperature ☐ Chloride ☐ Other	☐ Metals ☐ Pesticides/Herbicides ☐ Ammonia ☐ Dissolved Oxygen	☐ Nutrients ☐ Solids ☐ BOD
Status	☐ Completed & Adopted ☐ Completed & Rejected ☐ Under EPA review	☒Considered not developed ☐ Started - abandoned	☐On-going; ☐Under Consideration
Outcome	☐ Use Upgrade ☒Other: Considered but never developed	☐ Use Removal/Revision	☐ No Change in Use

Sulphur Creek UAA, Sandoval County, New Mexico

State(s)	NM	**EPA Region(s)** 6	
Leading effort	New Mexico Environmental Department, Surface Water Quality Bureau		
Contacts	Jane DeRose Bamman, New Mexico Environmental Department, Surface Water Quality Bureau		
Start	2000	**End**	On-going
Setting	☐ Effluent Dominated	☐ Wet-weather Impacted	☐ Urban
	☒ Other: Hot springs		
Pollutant Source Category	☐ Agricultural Runoff	☒ Natural Background	☐ Urban Stormwater
	☐ CSOs	☐ POTWs	☐ Mining
	☐ Industry	☐ Other:	
UAA Factor -131.10 (g)	1	☐ Undetermined	
Designated Use Assessed	☐ Aquatic Life	☒ Recreation	☒ Drinking Water
	☒ Other: Irrigation; fish culture		
Parameters of Concern	☐ Bacteria	☐ Metals	☐ Nutrients
	☐ Organic Hydrocarbons	☐ Pesticides/Herbicides	☐ Solids
	☐ Temperature	☐ Ammonia	☐ BOD
	☐ Chloride	☐ Dissolved Oxygen	
	☒ Other: pH		
Status	☐ Completed & Adopted	☐ Considered not developed	☐ On-going;
	☐ Completed & Rejected	☐ Started - abandoned	☐ Under Consideration
	☒ Under EPA review		
Outcome	☐ Use Upgrade	☐ Use Removal/Revision	☐ No Change in Use
	☒ Other: Under EPA review		

Christiana River UAA, Delaware

State(s)	DE	**EPA Region(s)** 3	
Leading effort	City of Wilmington		
Contacts	Sam Myoda, Delaware Department of Natural Resources and Environmental Control		
Start	2004	**End**	2005
Setting	☐ Effluent Dominated	☒ Wet-weather Impacted	☒ Urban
	☐ Other:		
Pollutant Source Category	☐ Agricultural Runoff	☐ Natural Background	☒ Urban Stormwater
	☒ CSOs	☐ POTWs	☐ Mining
	☐ Industry	☐ Other:	
UAA Factor -131.10 (g)		☒ Undetermined	
Designated Use Assessed	☐ Aquatic Life	☒ Recreation	☐ Drinking Water
	☐ Other		
Parameters of Concern	☒ Bacteria	☐ Metals	☐ Nutrients
	☐ Organic Hydrocarbons	☐ Pesticides/Herbicides	☐ Solids
	☐ Temperature	☐ Ammonia	☐ BOD
	☐ Chloride	☐ Dissolved Oxygen	
	☐ Other		
Status	☐ Completed & Adopted	☒Considered not developed	☐On-going;
	☐ Completed & Rejected	☐ Started - abandoned	☐Under Consideration
	☐ Under EPA review		
Outcome	☐ Use Upgrade	☐ Use Removal/Revision	☐ No Change in Use
	☒Other: Considered but never developed		

US Forest Service UAA for New World Mining District, Montana

State(s)	MT	**EPA Region(s)** 8	
Leading effort	US Forest Service		
Contacts	Bob Bukantis, Montana Department of Environmental Quality		
Start	NA	**End**	1999
Setting	☐ Effluent Dominated	☐ Wet-weather Impacted	☐ Urban
	☒ Other: Rural mine		
Pollutant Source Category	☐ Agricultural Runoff	☐ Natural Background	☐ Urban Stormwater
	☐ CSOs	☐ POTWs	☒ Mining
	☐ Industry	☐ Other:	
UAA Factor -131.10 (g)	3	☐ Undetermined	
Designated Use Assessed	☒ Aquatic Life	☒ Recreation	☒ Drinking Water
	☐ Other		
Parameters of Concern	☐ Bacteria	☒ Metals	☐ Nutrients
	☐ Organic Hydrocarbons	☐ Pesticides/Herbicides	☐ Solids
	☐ Temperature	☐ Ammonia	☐ BOD
	☐ Chloride	☐ Dissolved Oxygen	
	☐ Other		
Status	☐ Completed & Adopted	☒Considered not developed	☐On-going;
	☐ Completed & Rejected	☐ Started - abandoned	☐Under Consideration
	☐ Under EPA review		
Outcome	☐ Use Upgrade	☐ Use Removal/Revision	☒ No Change in Use
	☒Other: Temporary standards		

Denver International Airport UAA for South Platte River, Colorado

State(s)	CO	**EPA Region(s)** 8	
Leading effort	Denver International Airport		
Contacts	Eric Oppelt, Colorado Department of Public Health and Environment		
Start	2000	**End**	NA
Setting	☐ Effluent Dominated	☒ Wet-weather Impacted	☐ Urban
	☐ Other:		
Pollutant Source Category	☐ Agricultural Runoff	☐ Natural Background	☐ Urban Stormwater
	☐ CSOs	☐ POTWs	☐ Mining
	☐ Industry	☒ Other: Stormwater/de-icing runoff	
UAA Factor -131.10 (g)	2, 5	☐ Undetermined	
Designated Use Assessed	☒ Aquatic Life	☐ Recreation	☐ Drinking Water
	☐ Other		
Parameters of Concern	☐ Bacteria	☐ Metals	☐ Nutrients
	☐ Organic Hydrocarbons	☐ Pesticides/Herbicides	☐ Solids
	☐ Temperature	☐ Ammonia	☐ BOD
	☐ Chloride	☒ Dissolved Oxygen	
	☒ Other: Glycol		
Status	☐ Completed & Adopted	☒Considered not developed	☐On-going;
	☐ Completed & Rejected	☐ Started - abandoned	☐Under Consideration
	☐ Under EPA review		
Outcome	☐ Use Upgrade	☐ Use Removal/Revision	☒ No Change in Use
	☒Other: Site-specific criteria		

UAA for Harrisonburg WTP Chlorine Variance, Virginia

State(s)	VA	**EPA Region(s)** 3	
Leading effort	Virginia Department of Environmental Quality		
Contacts	David Whitehurst, Virginia Department of Environmental Quality		
Start	2000	**End**	2000
Setting	☒ Effluent Dominated	☐ Wet-weather Impacted	☐ Urban
	☐ Other:		
Pollutant Source Category	☐ Agricultural Runoff	☐ Natural Background	☐ Urban Stormwater
	☐ CSOs	☐ POTWs	☐ Mining
	☐ Industry	☒ Other: Water treatment plant discharge	
UAA Factor -131.10 (g)	2	☐ Undetermined	
Designated Use Assessed	☒ Aquatic Life	☐ Recreation	☐ Drinking Water
	☐ Other		
Parameters of Concern	☐ Bacteria	☐ Metals	☐ Nutrients
	☐ Organic Hydrocarbons	☐ Pesticides/Herbicides	☐ Solids
	☐ Temperature	☐ Ammonia	☐ BOD
	☐ Chloride	☐ Dissolved Oxygen	
	☒ Other: Chlorine		
Status	☐ Completed & Adopted	☐Considered not developed	☐On-going;
	☒ Completed & Rejected	☐ Started - abandoned	☐Under Consideration
	☐ Under EPA review		
Outcome	☐ Use Upgrade	☐ Use Removal/Revision	☒ No Change in Use
	☒Other: Approved by State; rejected by EPA		

Little Isaacs Creek UAA, Virginia

State(s)	VA	**EPA Region(s)** 3	
Leading effort	Virginia Department of Environmental Quality		
Contacts	David Whitehurst, Virginia Department of Environmental Quality		
Start	2001	**End**	2001
Setting	☒ Effluent Dominated	☐ Wet-weather Impacted	☐ Urban
	☐ Other:		
Pollutant Source Category	☐ Agricultural Runoff	☐ Natural Background	☐ Urban Stormwater
	☐ CSOs	☐ POTWs	☐ Mining
	☐ Industry	☒ Other: Private school STP; Seasonal <0.5 MGD	
UAA Factor -131.10 (g)	2	☐ Undetermined	
Designated Use Assessed	☒ Aquatic Life	☐ Recreation	☐ Drinking Water
	☐ Other		
Parameters of Concern	☐ Bacteria	☐ Metals	☐ Nutrients
	☐ Organic Hydrocarbons	☐ Pesticides/Herbicides	☐ Solids
	☐ Temperature	☒ Ammonia	☐ BOD
	☐ Chloride	☐ Dissolved Oxygen	
	☐ Other		
Status	☐ Completed & Adopted	☐Considered not developed	☐On-going;
	☒ Completed & Rejected	☐ Started - abandoned	☐Under Consideration
	☐ Under EPA review		
Outcome	☐ Use Upgrade	☐ Use Removal/Revision	☒ No Change in Use
	☒Other: Variance rejected by the State		

Lower French Gulch and Blue River UAA, Colorado

State(s)	CO	**EPA Region(s)** 8	
Leading effort	Summit Water Quality Committee, Silverthorne/Dillon JSA		
Contacts	Phil Hegeman, Colorado Department of Public Health and Environment		
Start	2000	**End**	2003
Setting	☒ Effluent Dominated	☐ Wet-weather Impacted	☐ Urban
	☒ Other: Rural Mine		
Pollutant Source Category	☐ Agricultural Runoff	☐ Natural Background	☐ Urban Stormwater
	☐ CSOs	☐ POTWs	☒ Mining
	☐ Industry	☐ Other:	
UAA Factor -131.10 (g)	5	☐ Undetermined	
Designated Use Assessed	☒ Aquatic Life	☐ Recreation	☐ Drinking Water
	☐ Other		
Parameters of Concern	☐ Bacteria	☒ Metals	☐ Nutrients
	☐ Organic Hydrocarbons	☐ Pesticides/Herbicides	☐ Solids
	☐ Temperature	☐ Ammonia	☐ BOD
	☐ Chloride	☐ Dissolved Oxygen	
	☐ Other		
Status	☒ Completed & Adopted	☐ Considered not developed	☐ On-going;
	☐ Completed & Rejected	☐ Started - abandoned	☐ Under Consideration
	☐ Under EPA review		
Outcome	☐ Use Upgrade	☒ Use Removal/Revision	☐ No Change in Use
	☒ Other: Site-specific criteria		

Numerous Recreational UAAs, Colorado

State(s)	CO	**EPA Region(s)** 8	
Leading effort	Streamside Landowner and Colorado River Water Conservation District		
Contacts	Phil Hegeman, Colorado Department of Public Health and Environment		
Start	2000	**End**	Ongoing
Setting	☐ Effluent Dominated	☐ Wet-weather Impacted	☐ Urban
	☒ Other: Rural low-flow streams		
Pollutant Source Category	☒ Agricultural Runoff	☐ Natural Background	☐ Urban Stormwater
	☐ CSOs	☐ POTWs	☐ Mining
	☐ Industry	☐ Other:	
UAA Factor -131.10 (g)	2	☐ Undetermined	
Designated Use Assessed	☐ Aquatic Life	☒ Recreation	☐ Drinking Water
	☐ Other		
Parameters of Concern	☒ Bacteria	☐ Metals	☐ Nutrients
	☐ Organic Hydrocarbons	☐ Pesticides/Herbicides	☐ Solids
	☐ Temperature	☐ Ammonia	☐ BOD
	☐ Chloride	☐ Dissolved Oxygen	
	☐ Other		
Status	☒ Completed & Adopted	☐ Considered not developed	☒ On-going;
	☐ Completed & Rejected	☐ Started - abandoned	☐ Under Consideration
	☐ Under EPA review		
Outcome	☐ Use Upgrade	☒ Use Removal/Revision	☒ No Change in Use
	☐ Other:		

Chicago Area Waterways System UAA, Illinois

State(s)	IL	**EPA Region(s)** 5	
Leading effort	Illinois Environmental Protection Agency		
Contacts	Rob Sulski, Illinois Environmental Protection Agency		
Start	2003	**End**	On-going
Setting	☒ Effluent Dominated	☒ Wet-weather Impacted	☒ Urban
	☐ Other:		
Pollutant Source Category	☐ Agricultural Runoff	☐ Natural Background	☒ Urban Stormwater
	☒ CSOs	☒ POTWs	☐ Mining
	☒ Industry	☐ Other:	
UAA Factor -131.10 (g)	1, 2, 3, 4, 5, 6	☐ Undetermined	
Designated Use Assessed	☒ Aquatic Life	☒ Recreation	☐ Drinking Water
	☐ Other		
Parameters of Concern	☒ Bacteria	☐ Metals	☐ Nutrients
	☐ Organic Hydrocarbons	☐ Pesticides/Herbicides	☐ Solids
	☒ Temperature	☐ Ammonia	☐ BOD
	☐ Chloride	☒ Dissolved Oxygen	
	☒ Other: Habitat		
Status	☐ Completed & Adopted	☐Considered not developed	☒On-going;
	☐ Completed & Rejected	☐ Started - abandoned	☐Under Consideration
	☐ Under EPA review		
Outcome	☐ Use Upgrade	☐ Use Removal/Revision	☐ No Change in Use
	☒Other: On-going		

Dell Arkansas Power Plant UAA for Drainage Ditches, Arkansas

State(s)	AR	**EPA Region(s)** 6	
Leading effort	FTN Associates, Ltd. for unnamed client		
Contacts	Jim Malcolm, FTN Associates, Ltd.		
Start	2005	**End**	2006
Setting	☒ Effluent Dominated	☐ Wet-weather Impacted	☐ Urban
	☐ Other:		
Pollutant Source Category	☒ Agricultural Runoff	☐ Natural Background	☐ Urban Stormwater
	☐ CSOs	☐ POTWs	☐ Mining
	☒ Industry	☐ Other:	
UAA Factor -131.10 (g)	1, 2, 3, 4, 5	☐ Undetermined	
Designated Use Assessed	☒ Aquatic Life	☒ Recreation	☒ Drinking Water
	☒ Other: Agricultural water supply		
Parameters of Concern	☐ Bacteria	☐ Metals	☐ Nutrients
	☐ Organic Hydrocarbons	☐ Pesticides/Herbicides	☐ Solids
	☒ Temperature	☐ Ammonia	☐ BOD
	☐ Chloride	☐ Dissolved Oxygen	
	☒ Other: Sulfate and TDS		
Status	☒ Completed & Adopted	☐Considered not developed	☐On-going;
	☐ Completed & Rejected	☐ Started - abandoned	☐Under Consideration
	☐ Under EPA review		
Outcome	☐ Use Upgrade	☒ Use Removal/Revision	☐ No Change in Use
	☒Other: Site specific criteria		

Marina Reclassifications for Shellfish Harvesting, North Carolina

State(s)	NC	**EPA Region(s)** 4	
Leading effort	North Carolina DENR		
Contacts	Bill Kreutzberger, CH2MHill		
Start	Mid 1980s	**End**	completed
Setting	☐ Effluent Dominated	☒ Wet-weather Impacted	☐ Urban
	☒ Other: Multiple bacteria sources		
Pollutant Source Category	☐ Agricultural Runoff	☐ Natural Background	☐ Urban Stormwater
	☐ CSOs	☐ POTWs	☐ Mining
	☐ Industry	☒ Other: Multiple	
UAA Factor -131.10 (g)	3	☐ Undetermined	
Designated Use Assessed	☐ Aquatic Life	☐ Recreation	☐ Drinking Water
	☒ Other: Shellfish harvesting		
Parameters of Concern	☒ Bacteria	☐ Metals	☐ Nutrients
	☐ Organic Hydrocarbons	☐ Pesticides/Herbicides	☐ Solids
	☐ Temperature	☐ Ammonia	☐ BOD
	☐ Chloride	☐ Dissolved Oxygen	
	☐ Other		
Status	☒ Completed & Adopted	☐ Considered not developed	☐ On-going;
	☐ Completed & Rejected	☐ Started - abandoned	☐ Under Consideration
	☐ Under EPA review		
Outcome	☐ Use Upgrade	☒ Use Removal/Revision	☐ No Change in Use
	☐ Other:		

Starkist UAA for Pago Pago Harbor, American Samoa

State(s)	America Samoa **EPA Region(s)** 9		
Leading effort	Starkist Tuna		
Contacts	Tom Dupuis, CH2MHill		
Start	2004	**End**	April 2006
Setting	☐ Effluent Dominated	☐ Wet-weather Impacted	☒ Urban
	☐ Other:		
Pollutant Source Category	☐ Agricultural Runoff	☐ Natural Background	☒ Urban Stormwater
	☐ CSOs	☒ POTWs	☐ Mining
	☒ Industry	☐ Other:	
UAA Factor -131.10 (g)	3	☐ Undetermined	
Designated Use Assessed	☒ Aquatic Life	☐ Recreation	☐ Drinking Water
	☐ Other		
Parameters of Concern	☐ Bacteria	☐ Metals	☒ Nutrients
	☐ Organic Hydrocarbons	☐ Pesticides/Herbicides	☐ Solids
	☐ Temperature	☐ Ammonia	☐ BOD
	☐ Chloride	☐ Dissolved Oxygen	
	☐ Other		
Status	☐ Completed & Adopted	☐Considered not developed	☐On-going;
	☒ Completed & Rejected	☐ Started - abandoned	☐Under Consideration
	☐ Under EPA review		
Outcome	☐ Use Upgrade	☐ Use Removal/Revision	☒ No Change in Use
	☐Other:		

Federal Paperboard UAA for Cape Fear River, North Carolina

State(s)	NC	**EPA Region(s)** 4	
Leading effort	Federal Paperboard		
Contacts	Bill Kreutzberger, CH2MHILL		
Start	1991	**End**	1993
Setting	☐ Effluent Dominated	☐ Wet-weather Impacted	☐ Urban
	☒ Other: Industrial NPDES		
Pollutant Source Category	☐ Agricultural Runoff	☐ Natural Background	☐ Urban Stormwater
	☐ CSOs	☐ POTWs	☐ Mining
	☒ Industry	☐ Other:	
UAA Factor -131.10 (g)	1	☐ Undetermined	
Designated Use Assessed	☒ Aquatic Life	☐ Recreation	☐ Drinking Water
	☐ Other		
Parameters of Concern	☐ Bacteria	☐ Metals	☐ Nutrients
	☐ Organic Hydrocarbons	☐ Pesticides/Herbicides	☐ Solids
	☐ Temperature	☐ Ammonia	☒ BOD
	☐ Chloride	☐ Dissolved Oxygen	
	☐ Other		
Status	☒ Completed & Adopted	☐Considered not developed	☐On-going;
	☐ Completed & Rejected	☐ Started - abandoned	☐Under Consideration
	☐ Under EPA review		
Outcome	☐ Use Upgrade	☐ Use Removal/Revision	☐ No Change in Use
	☒Other: Variance		

Lower Boise River Tributaries UAA, Idaho

State(s)	ID	**EPA Region(s)** 10	
Leading effort	Lower Boise Watershed Council		
Contacts	Tom Dupuis, CH2MHill		
Start	2000	**End**	2001
Setting	☐ Effluent Dominated	☐ Wet-weather Impacted	☒ Urban
	☐ Other:		
Pollutant Source Category	☒ Agricultural Runoff	☐ Natural Background	☐ Urban Stormwater
	☐ CSOs	☐ POTWs	☐ Mining
	☐ Industry	☐ Other:	
UAA Factor -131.10 (g)	2, 4, 5	☐ Undetermined	
Designated Use Assessed	☒ Aquatic Life	☒ Recreation	☐ Drinking Water
	☐ Other		
Parameters of Concern	☒ Bacteria	☐ Metals	☒ Nutrients
	☐ Organic Hydrocarbons	☐ Pesticides/Herbicides	☐ Solids
	☒ Temperature	☐ Ammonia	☐ BOD
	☐ Chloride	☐ Dissolved Oxygen	
	☒ Other: Sediment		
Status	☒ Completed & Adopted	☐Considered not developed	☐On-going;
	☒ Completed & Rejected	☐ Started - abandoned	☐Under Consideration
	☐ Under EPA review		
Outcome	☐ Use Upgrade	☐ Use Removal/Revision	☐ No Change in Use
	☒Other: On-going, EPA approved revised recreational uses but rejected revised aquatic life uses		

Bucktail Creek UAA, Idaho

State(s)	ID	**EPA Region(s)** 10	
Leading effort	Idaho DEQ		
Contacts	Chris Mebane, NOAA		
Start	2000	**End**	2002
Setting	☒ Effluent Dominated ☒ Other: Rural mine	☐ Wet-weather Impacted	☐ Urban
Pollutant Source Category	☐ Agricultural Runoff ☐ CSOs ☐ Industry	☐ Natural Background ☐ POTWs ☐ Other:	☐ Urban Stormwater ☒ Mining
UAA Factor -131.10 (g)	2, 3	☐ Undetermined	
Designated Use Assessed	☒ Aquatic Life ☐ Other	☒ Recreation	☐ Drinking Water
Parameters of Concern	☐ Bacteria ☐ Organic Hydrocarbons ☐ Temperature ☐ Chloride ☐ Other	☒ Metals ☐ Pesticides/Herbicides ☐ Ammonia ☐ Dissolved Oxygen	☐ Nutrients ☐ Solids ☐ BOD
Status	☒ Completed & Adopted ☐ Completed & Rejected ☐ Under EPA review	☐Considered not developed ☐ Started - abandoned	☐On-going; ☐Under Consideration
Outcome	☐ Use Upgrade ☐Other:	☒ Use Removal/Revision	☐ No Change in Use

City of Quincy UAA for Wasteways, Washington

State(s)	WA	**EPA Region(s)** 10	
Leading effort	City of Quincy		
Contacts	Phillip Burke		
Start	1998	**End**	NA
Setting	☒ Effluent Dominated	☐ Wet-weather Impacted	☐ Urban
	☐ Other:		
Pollutant Source Category	☐ Agricultural Runoff	☐ Natural Background	☐ Urban Stormwater
	☐ CSOs	☒ POTWs	☐ Mining
	☐ Industry	☐ Other:	
UAA Factor -131.10 (g)	3	☐ Undetermined	
Designated Use Assessed	☒ Aquatic Life	☐ Recreation	☐ Drinking Water
	☐ Other		
Parameters of Concern	☒ Bacteria	☐ Metals	☒ Nutrients
	☐ Organic Hydrocarbons	☐ Pesticides/Herbicides	☒ Solids
	☒ Temperature	☐ Ammonia	☐ BOD
	☐ Chloride	☐ Dissolved Oxygen	
	☐ Other		
Status	☐ Completed & Adopted	☐Considered not developed	☐On-going;
	☐ Completed & Rejected	☒ Started - abandoned	☐Under Consideration
	☐ Under EPA review		
Outcome	☐ Use Upgrade	☐ Use Removal/Revision	☒ No Change in Use
	☐Other:		

Santa Ana River UAA, California

State(s)	CA	**EPA Region(s)** 9	
Leading effort	Multiple NPDES MS4 Permittees		
Contacts	Tim White		
Start	1989	**End**	1994
Setting	☒ Effluent Dominated	☒ Wet-weather Impacted	☒ Urban
	☐ Other:		
Pollutant Source Category	☐ Agricultural Runoff	☐ Natural Background	☒ Urban Stormwater
	☐ CSOs	☒ POTWs	☐ Mining
	☒ Industry	☐ Other:	
UAA Factor -131.10 (g)	6	☐ Undetermined	
Designated Use Assessed	☒ Aquatic Life	☐ Recreation	☐ Drinking Water
	☐ Other		
Parameters of Concern	☐ Bacteria	☒ Metals	☐ Nutrients
	☐ Organic Hydrocarbons	☐ Pesticides/Herbicides	☐ Solids
	☐ Temperature	☒ Ammonia	☐ BOD
	☒ Chloride	☐ Dissolved Oxygen	
	☐ Other		
Status	☒ Completed & Adopted	☐ Considered not developed	☐ On-going;
	☐ Completed & Rejected	☐ Started - abandoned	☐ Under Consideration
	☐ Under EPA review		
Outcome	☐ Use Upgrade	☒ Use Removal/Revision	☐ No Change in Use
	☐ Other:		

Marathon Oil Company UAA for Coal Draw, Wyoming

State(s)	WY	**EPA Region(s)** 8	
Leading effort	Marathon Oil Company		
Contacts	Bill DeRienzo, Wyoming DEQ		
Start	2001	**End**	2002
Setting	☒ Effluent Dominated	☐ Wet-weather Impacted	☐ Urban
	☐ Other:		
Pollutant Source Category	☐ Agricultural Runoff	☐ Natural Background	☐ Urban Stormwater
	☐ CSOs	☐ POTWs	☐ Mining
	☒ Industry	☐ Other:	
UAA Factor -131.10 (g)	3	☐ Undetermined	
Designated Use Assessed	☒ Aquatic Life	☐ Recreation	☐ Drinking Water
	☐ Other		
Parameters of Concern	☐ Bacteria	☐ Metals	☒ Nutrients
	☐ Organic Hydrocarbons	☐ Pesticides/Herbicides	☐ Solids
	☒ Temperature	☐ Ammonia	☐ BOD
	☒ Chloride	☐ Dissolved Oxygen	
	☐ Other		
Status	☐ Completed & Adopted	☐Considered not developed	☐On-going;
	☒ Completed & Rejected	☐ Started - abandoned	☐Under Consideration
	☐ Under EPA review		
Outcome	☐ Use Upgrade	☒ Use Removal/Revision	☐ No Change in Use
	☐Other:		

Class 4C Isolated Effluent Statewide UAA, Wyoming

State(s)	WY	**EPA Region(s)** 8	
Leading effort	Wyoming DEQ		
Contacts	Bill DiRienzo, Wyoming DEQ		
Start	2001	**End**	2002
Setting	☒ Effluent Dominated	☐ Wet-weather Impacted	☐ Urban
	☐ Other:		
Pollutant Source Category	☐ Agricultural Runoff	☒ Natural Background	☐ Urban Stormwater
	☐ CSOs	☐ POTWs	☐ Mining
	☐ Industry	☐ Other:	
UAA Factor -131.10 (g)	3	☐ Undetermined	
Designated Use Assessed	☒ Aquatic Life	☐ Recreation	☐ Drinking Water
	☐ Other		
Parameters of Concern	☐ Bacteria	☐ Metals	☒ Nutrients
	☐ Organic Hydrocarbons	☐ Pesticides/Herbicides	☐ Solids
	☒ Temperature	☐ Ammonia	☐ BOD
	☐ Chloride	☐ Dissolved Oxygen	
	☐ Other		
Status	☒ Completed & Adopted	☐ Considered not developed	☐ On-going;
	☐ Completed & Rejected	☐ Started - abandoned	☐ Under Consideration
	☐ Under EPA review		
Outcome	☐ Use Upgrade	☒ Use Removal/Revision	☐ No Change in Use
	☐ Other:		

Class 4A Artificially Constructed Canals and Ditches Statewide UAA, Wyoming

State(s)	WY	**EPA Region(s)** 8	
Leading effort	Wyoming DEQ		
Contacts	Bill DiRienzo, Wyoming DEQ		
Start	2002	**End**	2005
Setting	☒ Effluent Dominated	☐ Wet-weather Impacted	☐ Urban
	☐ Other:		
Pollutant Source Category	☒ Agricultural Runoff	☐ Natural Background	☐ Urban Stormwater
	☐ CSOs	☐ POTWs	☐ Mining
	☐ Industry	☐ Other:	
UAA Factor -131.10 (g)	5	☐ Undetermined	
Designated Use Assessed	☒ Aquatic Life	☐ Recreation	☐ Drinking Water
	☐ Other		
Parameters of Concern	☒ Bacteria	☐ Metals	☒ Nutrients
	☐ Organic Hydrocarbons	☐ Pesticides/Herbicides	☒ Solids
	☒ Temperature	☐ Ammonia	☐ BOD
	☐ Chloride	☐ Dissolved Oxygen	
	☐ Other		
Status	☒ Completed & Adopted	☐ Considered not developed	☐ On-going;
	☐ Completed & Rejected	☐ Started - abandoned	☐ Under Consideration
	☐ Under EPA review		
Outcome	☐ Use Upgrade	☒ Use Removal/Revision	☐ No Change in Use
	☐ Other:		

Blackbird Creek UAA, Idaho

State(s)	ID	**EPA Region(s)** 10	
Leading effort	Idaho DEQ		
Contacts	Chris Mebane, NOAA		
Start	1996	**End**	1998
Setting	☒ Effluent Dominated ☒ Other: Rural mine	☐ Wet-weather Impacted	☐ Urban
Pollutant Source Category	☐ Agricultural Runoff ☐ CSOs ☐ Industry	☐ Natural Background ☐ POTWs ☐ Other:	☐ Urban Stormwater ☒ Mining
UAA Factor -131.10 (g)	3	☐ Undetermined	
Designated Use Assessed	☒ Aquatic Life ☐ Other	☐ Recreation	☐ Drinking Water
Parameters of Concern	☐ Bacteria ☐ Organic Hydrocarbons ☐ Temperature ☐ Chloride ☐ Other	☒ Metals ☐ Pesticides/Herbicides ☐ Ammonia ☐ Dissolved Oxygen	☐ Nutrients ☐ Solids ☐ BOD
Status	☒ Completed & Adopted ☐ Completed & Rejected ☐ Under EPA review	☐Considered not developed ☐ Started - abandoned	☐On-going; ☐Under Consideration
Outcome	☐ Use Upgrade ☐Other:	☒ Use Removal/Revision	☐ No Change in Use

Spokane River and Long Lake Reservoir UAA, Idaho and Washington

State(s)	ID, WA	**EPA Region(s)** 10	
Leading effort	Municipal and industrial dischargers		
Contacts	Dale Arnold, City of Spokane		
Start	2003	**End**	2007 expected
Setting	☒ Effluent Dominated	☒ Wet-weather Impacted	☒ Urban
	☐ Other:		
Pollutant Source Category	☐ Agricultural Runoff	☐ Natural Background	☐ Urban Stormwater
	☐ CSOs	☒ POTWs	☐ Mining
	☒ Industry	☐ Other:	
UAA Factor -131.10 (g)	4, 5	☐ Undetermined	
Designated Use Assessed	☒ Aquatic Life	☐ Recreation	☐ Drinking Water
	☐ Other		
Parameters of Concern	☐ Bacteria	☐ Metals	☒ Nutrients
	☐ Organic Hydrocarbons	☐ Pesticides/Herbicides	☐ Solids
	☒ Temperature	☐ Ammonia	☐ BOD
	☐ Chloride	☒ Dissolved Oxygen	
	☐ Other		
Status	☐ Completed & Adopted	☐Considered not developed	☐On-going;
	☐ Completed & Rejected	☒ Started - abandoned	☐Under Consideration
	☐ Under EPA review		
Outcome	☐ Use Upgrade	☐ Use Removal/Revision	☒ No Change in Use
	☐Other: On-going		

Donahue Industries UAA for Angelina River, Texas

State(s)	TX	**EPA Region(s)** 8	
Leading effort	Donohue Industries		
Contacts	Michael Bastian		
Start	1994	**End**	2000
Setting	☒ Effluent Dominated	☐ Wet-weather Impacted	☐ Urban
	☐ Other:		
Pollutant Source Category	☐ Agricultural Runoff	☐ Natural Background	☐ Urban Stormwater
	☐ CSOs	☐ POTWs	☐ Mining
	☒ Industry	☐ Other:	
UAA Factor -131.10 (g)	2, 3	☐ Undetermined	
Designated Use Assessed	☒ Aquatic Life	☐ Recreation	☐ Drinking Water
	☐ Other		
Parameters of Concern	☐ Bacteria	☐ Metals	☐ Nutrients
	☐ Organic Hydrocarbons	☐ Pesticides/Herbicides	☐ Solids
	☐ Temperature	☐ Ammonia	☐ BOD
	☐ Chloride	☒ Dissolved Oxygen	
	☐ Other		
Status	☐ Completed & Adopted	☐ Considered not developed	☐ On-going;
	☒ Completed & Rejected	☐ Started - abandoned	☐ Under Consideration
	☐ Under EPA review		
Outcome	☐ Use Upgrade	☐ Use Removal/Revision	☒ No Change in Use
	☐ Other:		

Malheur River UAA, Oregon

State(s)	OR	**EPA Region(s)** 10	
Leading effort	US EPA		
Contacts	NA		
Start	1986	**End**	1986
Setting	☒ Effluent Dominated ☐ Other:	☐ Wet-weather Impacted	☐ Urban
Pollutant Source Category	☒ Agricultural Runoff ☐ CSOs ☐ Industry	☐ Natural Background ☐ POTWs ☐ Other:	☐ Urban Stormwater ☐ Mining
UAA Factor -131.10 (g)	1, 3, 4	☐ Undetermined	
Designated Use Assessed	☒ Aquatic Life ☐ Other	☐ Recreation	☐ Drinking Water
Parameters of Concern	☐ Bacteria ☐ Organic Hydrocarbons ☒ Temperature ☐ Chloride ☐ Other	☐ Metals ☐ Pesticides/Herbicides ☐ Ammonia ☐ Dissolved Oxygen	☒ Nutrients ☒ Solids ☐ BOD
Status	☒ Completed & Adopted ☐ Completed & Rejected ☐ Under EPA review	☐Considered not developed ☐ Started - abandoned	☐On-going; ☐Under Consideration
Outcome	☐ Use Upgrade ☐Other:	☒ Use Removal/Revision	☐ No Change in Use

Engineered Flood Channels UAA for Recreational Uses in Wet Weather, California

State(s)	CA	**EPA Region(s)** 9	
Leading effort	Regional WQ Control Board, Los Angeles Region		
Contacts	Tim Moore, Risk Sciences		
Start	Late 1990's	**End**	2003
Setting	☐ Effluent Dominated ☐ Other:	☒ Wet-weather Impacted	☒ Urban
Pollutant Source Category	☐ Agricultural Runoff ☐ CSOs ☐ Industry	☐ Natural Background ☐ POTWs ☐ Other:	☒ Urban Stormwater ☐ Mining
UAA Factor -131.10 (g)	3	☐ Undetermined	
Designated Use Assessed	☐ Aquatic Life ☐ Other	☒ Recreation	☐ Drinking Water
Parameters of Concern	☒ Bacteria ☐ Organic Hydrocarbons ☐ Temperature ☐ Chloride ☐ Other	☐ Metals ☐ Pesticides/Herbicides ☐ Ammonia ☐ Dissolved Oxygen	☐ Nutrients ☐ Solids ☐ BOD
Status	☒ Completed & Adopted ☐ Completed & Rejected ☐ Under EPA review	☐Considered not developed ☐ Started - abandoned	☐On-going; ☐Under Consideration
Outcome	☐ Use Upgrade ☐Other:	☒ Use Removal/Revision	☐ No Change in Use

Statewide Recreational UAAs, Kansas

State(s)	KS	**EPA Region(s)** 7	
Leading effort	Kansas DHE		
Contacts	Craig Thompson, KDHE		
Start	1998	**End**	2006
Setting	☒ Effluent Dominated	☒ Wet-weather Impacted	☒ Urban
	☐ Other:		
Pollutant Source Category	☒ Agricultural Runoff	☒ Natural Background	☒ Urban Stormwater
	☐ CSOs	☒ POTWs	☐ Mining
	☐ Industry	☐ Other:	
UAA Factor -131.10 (g)	2, 5	☐ Undetermined	
Designated Use Assessed	☐ Aquatic Life	☒ Recreation	☐ Drinking Water
	☐ Other		
Parameters of Concern	☒ Bacteria	☐ Metals	☐ Nutrients
	☐ Organic Hydrocarbons	☐ Pesticides/Herbicides	☐ Solids
	☐ Temperature	☐ Ammonia	☐ BOD
	☐ Chloride	☐ Dissolved Oxygen	
	☐ Other		
Status	☒ Completed & Adopted	☐Considered not developed	☒On-going;
	☐ Completed & Rejected	☐ Started - abandoned	☐Under Consideration
	☐ Under EPA review		
Outcome	☒ Use Upgrade	☒ Use Removal/Revision	☒ No Change in Use
	☐Other:		

Snake River-Hells Canyon UAA, Idaho and Oregon

State(s)	ID, OR	**EPA Region(s)** 10	
Leading effort	Local stakeholders		
Contacts	Don Essig		
Start	2003	**End**	2008
Setting	☐ Effluent Dominated	☐ Wet-weather Impacted	☒ Urban
	☐ Other:		
Pollutant Source Category	☒ Agricultural Runoff	☐ Natural Background	☐ Urban Stormwater
	☐ CSOs	☒ POTWs	☐ Mining
	☒ Industry	☐ Other:	
UAA Factor -131.10 (g)	1, 3, 4	☐ Undetermined	
Designated Use Assessed	☒ Aquatic Life	☐ Recreation	☐ Drinking Water
	☐ Other		
Parameters of Concern	☐ Bacteria	☐ Metals	☒ Nutrients
	☐ Organic Hydrocarbons	☐ Pesticides/Herbicides	☐ Solids
	☒ Temperature	☐ Ammonia	☐ BOD
	☐ Chloride	☐ Dissolved Oxygen	
	☐ Other		
Status	☐ Completed & Adopted	☐ Considered not developed	☒ On-going;
	☐ Completed & Rejected	☐ Started - abandoned	☐ Under Consideration
	☐ Under EPA review		
Outcome	☐ Use Upgrade	☐ Use Removal/Revision	☐ No Change in Use
	☒ Other: On-going		

Boston Harbor UAA for CSO Discharges, Massachusetts

State(s)	MA	**EPA Region(s)** 1	
Leading effort	Massachusetts Water Resources Authority, MA DEP		
Contacts	Glenn Haas; Kevin Brander		
Start	1992	**End**	1997
Setting	☐ Effluent Dominated	☒ Wet-weather Impacted	☒ Urban
	☐ Other:		
Pollutant Source Category	☐ Agricultural Runoff	☐ Natural Background	☐ Urban Stormwater
	☒ CSOs	☐ POTWs	☐ Mining
	☐ Industry	☐ Other:	
UAA Factor -131.10 (g)	6	☐ Undetermined	
Designated Use Assessed	☐ Aquatic Life	☒ Recreation	☐ Drinking Water
	☐ Other		
Parameters of Concern	☒ Bacteria	☐ Metals	☐ Nutrients
	☐ Organic Hydrocarbons	☐ Pesticides/Herbicides	☐ Solids
	☐ Temperature	☐ Ammonia	☐ BOD
	☐ Chloride	☐ Dissolved Oxygen	
	☐ Other		
Status	☒ Completed & Adopted	☐ Considered not developed	☐ On-going;
	☐ Completed & Rejected	☐ Started - abandoned	☐ Under Consideration
	☐ Under EPA review		
Outcome	☐ Use Upgrade	☒ Use Removal/Revision	☐ No Change in Use
	☒ Other: Variance		

Statewide Aquatic Life UAAs, Ohio

State(s)	OH	**EPA Region(s)** 5	
Leading effort	Ohio EPA		
Contacts	Chris Yoder		
Start	evolving since 1970s	**End**	On-going
Setting	☒ Effluent Dominated ☐ Other:	☒ Wet-weather Impacted	☐ Urban
Pollutant Source Category	☒ Agricultural Runoff ☐ CSOs ☐ Industry	☒ Natural Background ☐ POTWs ☐ Other:	☐ Urban Stormwater ☐ Mining
UAA Factor -131.10 (g)	1, 2, 3, 4, 5	☐ Undetermined	
Designated Use Assessed	☒ Aquatic Life ☐ Other	☒ Recreation	☒ Drinking Water
Parameters of Concern	☐ Bacteria ☐ Organic Hydrocarbons ☐ Temperature ☐ Chloride ☒ Other: Biological assessment	☐ Metals ☐ Pesticides/Herbicides ☐ Ammonia ☐ Dissolved Oxygen	☐ Nutrients ☐ Solids ☐ BOD
Status	☒ Completed & Adopted ☐ Completed & Rejected ☐ Under EPA review	☐ Considered not developed ☐ Started - abandoned	☐ On-going; ☐ Under Consideration
Outcome	☒ Use Upgrade ☐ Other:	☒ Use Removal/Revision	☒ No Change in Use

Statewide Aquatic Life UAAs, Oklahoma

State(s)	OK	**EPA Region(s)** 6	
Leading effort	Oklahoma State Water Board		
Contacts	Chuck Potts		
Start	1980's	**End**	Ongoing
Setting	☒ Effluent Dominated	☒ Wet-weather Impacted	☐ Urban
	☐ Other:		
Pollutant Source Category	☐ Agricultural Runoff	☐ Natural Background	☐ Urban Stormwater
	☐ CSOs	☒ POTWs	☐ Mining
	☐ Industry	☐ Other:	
UAA Factor -131.10 (g)	2	☐ Undetermined	
Designated Use Assessed	☒ Aquatic Life	☒ Recreation	☒ Drinking Water
	☐ Other		
Parameters of Concern	☐ Bacteria	☐ Metals	☐ Nutrients
	☐ Organic Hydrocarbons	☐ Pesticides/Herbicides	☐ Solids
	☐ Temperature	☐ Ammonia	☐ BOD
	☐ Chloride	☐ Dissolved Oxygen	
	☒ Other: Varies		
Status	☒ Completed & Adopted	☐ Considered not developed	☐ On-going;
	☐ Completed & Rejected	☐ Started - abandoned	☐ Under Consideration
	☐ Under EPA review		
Outcome	☒ Use Upgrade	☒ Use Removal/Revision	☒ No Change in Use
	☐ Other:		

Cabin Creek UAA, Georgia

State(s)	GA	**EPA Region(s)** 4	
Leading effort	EPA Region 4		
Contacts	Eve Zimmerman, EPA Region 4		
Start	NA	**End**	NA
Setting	☒ Effluent Dominated	☐ Wet-weather Impacted	☐ Urban
	☐ Other:		
Pollutant Source Category	☐ Agricultural Runoff	☐ Natural Background	☐ Urban Stormwater
	☐ CSOs	☐ POTWs	☐ Mining
	☒ Industry	☐ Other:	
UAA Factor -131.10 (g)	6	☐ Undetermined	
Designated Use Assessed	☒ Aquatic Life	☐ Recreation	☐ Drinking Water
	☐ Other		
Parameters of Concern	☐ Bacteria	☐ Metals	☐ Nutrients
	☐ Organic Hydrocarbons	☐ Pesticides/Herbicides	☐ Solids
	☐ Temperature	☐ Ammonia	☐ BOD
	☒ Chloride	☐ Dissolved Oxygen	
	☐ Other		
Status	☒ Completed & Adopted	☐ Considered not developed	☐ On-going;
	☐ Completed & Rejected	☐ Started - abandoned	☐ Under Consideration
	☐ Under EPA review		
Outcome	☐ Use Upgrade	☐ Use Removal/Revision	☐ No Change in Use
	☒ Other: Variance		

Squaw Creek UAA, California

State(s)	CA	**EPA Region(s)** 9	
Leading effort	CA Regional Water Quality Control Board, Region 5		
Contacts	Phil Woodward		
Start	2002	**End**	late 2006
Setting	☐ Effluent Dominated	☐ Wet-weather Impacted	☐ Urban
	☒ Other:		
Pollutant Source Category	☐ Agricultural Runoff	☐ Natural Background	☐ Urban Stormwater
	☐ CSOs	☐ POTWs	☒ Mining
	☐ Industry	☐ Other:	
UAA Factor -131.10 (g)	3	☐ Undetermined	
Designated Use Assessed	☒ Aquatic Life	☐ Recreation	☐ Drinking Water
	☐ Other		
Parameters of Concern	☐ Bacteria	☒ Metals	☐ Nutrients
	☐ Organic Hydrocarbons	☐ Pesticides/Herbicides	☐ Solids
	☐ Temperature	☐ Ammonia	☐ BOD
	☐ Chloride	☐ Dissolved Oxygen	
	☐ Other		
Status	☐ Completed & Adopted	☐ Considered not developed	☐ On-going;
	☐ Completed & Rejected	☐ Started - abandoned	☒ Under Consideration
	☐ Under EPA review		
Outcome	☐ Use Upgrade	☐ Use Removal/Revision	☐ No Change in Use
	☒ Other: On-going		

Fenholloway River UAA, Florida

State(s)	FL	**EPA Region(s)** 4	
Leading effort	Florida DEQ		
Contacts	Greg Knecht, FLA DEP		
Start	1992	**End**	1994
Setting	☐ Effluent Dominated	☐ Wet-weather Impacted	☐ Urban
	☒ Other:		
Pollutant Source Category	☐ Agricultural Runoff	☐ Natural Background	☐ Urban Stormwater
	☐ CSOs	☐ POTWs	☐ Mining
	☒ Industry	☐ Other:	
UAA Factor -131.10 (g)	6	☐ Undetermined	
Designated Use Assessed	☒ Aquatic Life	☒ Recreation	☐ Drinking Water
	☐ Other		
Parameters of Concern	☐ Bacteria	☐ Metals	☒ Nutrients
	☐ Organic Hydrocarbons	☐ Pesticides/Herbicides	☐ Solids
	☐ Temperature	☒ Ammonia	☒ BOD
	☐ Chloride	☒ Dissolved Oxygen	
	☒ Other: Color, conductivity		
Status	☒ Completed & Adopted	☐ Considered not developed	☐ On-going;
	☐ Completed & Rejected	☐ Started - abandoned	☐ Under Consideration
	☐ Under EPA review		
Outcome	☒ Use Upgrade	☐ Use Removal/Revision	☐ No Change in Use
	☐ Other:		

New Iberia UAA, Louisiana

State(s)	LA	**EPA Region(s)** 6	
Leading effort	Louisiana DEQ		
Contacts	Kris Pintado, LA DEQ		
Start	2003	**End**	2004
Setting	☐ Effluent Dominated	☒ Wet-weather Impacted	☒ Urban
	☐ Other:		
Pollutant Source Category	☒ Agricultural Runoff	☐ Natural Background	☒ Urban Stormwater
	☐ CSOs	☒ POTWs	☐ Mining
	☐ Industry	☐ Other:	
UAA Factor -131.10 (g)	3, 4	☐ Undetermined	
Designated Use Assessed	☒ Aquatic Life	☐ Recreation	☐ Drinking Water
	☐ Other		
Parameters of Concern	☐ Bacteria	☐ Metals	☐ Nutrients
	☐ Organic Hydrocarbons	☐ Pesticides/Herbicides	☐ Solids
	☐ Temperature	☐ Ammonia	☐ BOD
	☐ Chloride	☒ Dissolved Oxygen	
	☐ Other		
Status	☒ Completed & Adopted	☐ Considered not developed	☐ On-going;
	☐ Completed & Rejected	☐ Started - abandoned	☐ Under Consideration
	☐ Under EPA review		
Outcome	☐ Use Upgrade	☒ Use Removal/Revision	☐ No Change in Use
	☒ Other: Use refinement		

Owen's Lake UAA, California

State(s)	CA	**EPA Region(s)** 9	
Leading effort	Lahontan Regional Water Quality Control Board		
Contacts	Judith Unsicker, Lahontan RWQCB		
Start	2004	**End**	2006
Setting	☐ Effluent Dominated	☐ Wet-weather Impacted	☐ Urban
	☒ Other:		
Pollutant Source Category	☐ Agricultural Runoff	☒ Natural Background	☐ Urban Stormwater
	☐ CSOs	☐ POTWs	☐ Mining
	☐ Industry	☐ Other:	
UAA Factor -131.10 (g)	1, 4	☐ Undetermined	
Designated Use Assessed	☐ Aquatic Life	☐ Recreation	☒ Drinking Water
	☐ Other		
Parameters of Concern	☐ Bacteria	☐ Metals	☐ Nutrients
	☐ Organic Hydrocarbons	☐ Pesticides/Herbicides	☐ Solids
	☐ Temperature	☐ Ammonia	☐ BOD
	☐ Chloride	☐ Dissolved Oxygen	
	☒ Other: TDS		
Status	☐ Completed & Adopted	☐ Considered not developed	☒ On-going;
	☐ Completed & Rejected	☐ Started - abandoned	☐ Under Consideration
	☐ Under EPA review		
Outcome	☐ Use Upgrade	☒ Use Removal/Revision	☐ No Change in Use
	☐ Other:		

City of Vacaville UAA for Old Alamo Creek, California

State(s)	CA	**EPA Region(s)** 9	
Leading effort	City of Vacaville		
Contacts	Jacque McCall, City of Vacaville		
Start	2001	**End**	2006
Setting	☒ Effluent Dominated	☐ Wet-weather Impacted	☐ Urban
	☐ Other:		
Pollutant Source Category	☒ Agricultural Runoff	☐ Natural Background	☐ Urban Stormwater
	☐ CSOs	☒ POTWs	☐ Mining
	☐ Industry	☐ Other:	
UAA Factor -131.10 (g)	2, 3	☐ Undetermined	
Designated Use Assessed	☒ Aquatic Life	☐ Recreation	☒ Drinking Water
	☐ Other		
Parameters of Concern	☐ Bacteria	☐ Metals	☐ Nutrients
	☐ Organic Hydrocarbons	☐ Pesticides/Herbicides	☐ Solids
	☐ Temperature	☐ Ammonia	☐ BOD
	☐ Chloride	☐ Dissolved Oxygen	
	☒ Other: Trihalomethanes		
Status	☐ Completed & Adopted	☐Considered not developed	☒On-going;
	☐ Completed & Rejected	☐ Started - abandoned	☐Under Consideration
	☐ Under EPA review		
Outcome	☐ Use Upgrade	☒ Use Removal/Revision	☐ No Change in Use
	☐Other:		

City of San Luis Obispo UAA for San Luis Obispo Creek, California

State(s)	CA	**EPA Region(s)** 9	
Leading effort	City of San Luis Obisbo		
Contacts	Dave Hix, City of San Luis Obisbo		
Start	2005	**End**	2006
Setting	☒ Effluent Dominated	☐ Wet-weather Impacted	☐ Urban
	☐ Other:		
Pollutant Source Category	☐ Agricultural Runoff	☐ Natural Background	☐ Urban Stormwater
	☐ CSOs	☒ POTWs	☐ Mining
	☐ Industry	☐ Other:	
UAA Factor -131.10 (g)	2	☐ Undetermined	
Designated Use Assessed	☐ Aquatic Life	☐ Recreation	☒ Drinking Water
	☐ Other		
Parameters of Concern	☐ Bacteria	☐ Metals	☒ Nutrients
	☐ Organic Hydrocarbons	☐ Pesticides/Herbicides	☐ Solids
	☐ Temperature	☐ Ammonia	☐ BOD
	☐ Chloride	☐ Dissolved Oxygen	
	☐ Other		
Status	☐ Completed & Adopted	☐Considered not developed	☒On-going;
	☐ Completed & Rejected	☐ Started - abandoned	☐Under Consideration
	☐ Under EPA review		
Outcome	☐ Use Upgrade	☒ Use Removal/Revision	☐ No Change in Use
	☐Other:		

Naturally Impaired Waters of the Lahontan Region Use Removal, California

State(s)	Ca	**EPA Region(s)** 9	
Leading effort	Lahontan Regional Water Quality Control Board		
Contacts	Judith Unsicker, RWQCB		
Start	1998/1999	**End**	2002
Setting	☐ Effluent Dominated	☐ Wet-weather Impacted	☐ Urban
	☒ Other:		
Pollutant Source Category	☐ Agricultural Runoff	☒ Natural Background	☐ Urban Stormwater
	☐ CSOs	☐ POTWs	☐ Mining
	☐ Industry	☐ Other:	
UAA Factor -131.10 (g)	1	☐ Undetermined	
Designated Use Assessed	☐ Aquatic Life	☐ Recreation	☒ Drinking Water
	☐ Other		
Parameters of Concern	☐ Bacteria	☐ Metals	☐ Nutrients
	☐ Organic Hydrocarbons	☐ Pesticides/Herbicides	☐ Solids
	☐ Temperature	☐ Ammonia	☐ BOD
	☐ Chloride	☐ Dissolved Oxygen	
	☒ Other: Salt; trace elements (arsenic, boron, fluoride)		
Status	☒ Completed & Adopted	☐ Considered not developed	☐ On-going;
	☐ Completed & Rejected	☐ Started - abandoned	☐ Under Consideration
	☐ Under EPA review		
Outcome	☐ Use Upgrade	☒ Use Removal/Revision	☐ No Change in Use
	☐ Other:		

East LA Lancaster POTW UAA, California

State(s)	CA	**EPA Region(s)** 9	
Leading effort	Lahontan Regional Water Quality Control Board		
Contacts	Judith Unsicker, RWQCB		
Start	2002	**End**	2007
Setting	☐ Effluent Dominated	☐ Wet-weather Impacted	☐ Urban
	☒ Other:		
Pollutant Source Category	☐ Agricultural Runoff	☐ Natural Background	☐ Urban Stormwater
	☐ CSOs	☒ POTWs	☐ Mining
	☐ Industry	☐ Other:	
UAA Factor -131.10 (g)		☒ Undetermined	
Designated Use Assessed	☐ Aquatic Life	☒ Recreation	☒ Drinking Water
	☐ Other		
Parameters of Concern	☐ Bacteria	☐ Metals	☐ Nutrients
	☐ Organic Hydrocarbons	☐ Pesticides/Herbicides	☐ Solids
	☐ Temperature	☐ Ammonia	☐ BOD
	☐ Chloride	☐ Dissolved Oxygen	
	☒ Other: Chlorine; pH		
Status	☐ Completed & Adopted	☐ Considered not developed	☒ On-going;
	☐ Completed & Rejected	☐ Started - abandoned	☐ Under Consideration
	☐ Under EPA review		
Outcome	☐ Use Upgrade	☒ Use Removal/Revision	☐ No Change in Use
	☐ Other:		

Central Valley Selenium and Boron UAA, California

State(s)	CA	**EPA Region(s)** 9	
Leading effort	Central Valley Regional Water Quality Control Board		
Contacts	Rudy Shnagle, RWQCB		
Start	1990	**End**	1996
Setting	☒ Effluent Dominated ☐ Other:	☐ Wet-weather Impacted	☐ Urban
Pollutant Source Category	☐ Agricultural Runoff ☐ CSOs ☐ Industry	☒ Natural Background ☐ POTWs ☐ Other:	☐ Urban Stormwater ☐ Mining
UAA Factor -131.10 (g)	1, 6	☐ Undetermined	
Designated Use Assessed	☐ Aquatic Life ☐ Other	☒ Recreation	☐ Drinking Water
Parameters of Concern	☐ Bacteria ☐ Organic Hydrocarbons ☐ Temperature ☐ Chloride ☒ Other: Selenium and boron	☐ Metals ☐ Pesticides/Herbicides ☐ Ammonia ☐ Dissolved Oxygen	☐ Nutrients ☐ Solids ☐ BOD
Status	☐ Completed & Adopted ☒ Completed & Rejected ☐ Under EPA review	☐Considered not developed ☐ Started - abandoned	☐On-going; ☐Under Consideration
Outcome	☐ Use Upgrade ☐Other:	☐ Use Removal/Revision	☒ No Change in Use

City of Vacaville UAA for New Alamo Creek, Cache Slough, and Ulatis Creek, California

State(s)	CA	**EPA Region(s)** 9	
Leading effort	City of Vacaville, Central Valley California		
Contacts	Jacque McCall, City of Vacaville		
Start	2004	**End**	2008
Setting	☒ Effluent Dominated ☐ Other:	☐ Wet-weather Impacted	☐ Urban
Pollutant Source Category	☒ Agricultural Runoff ☐ CSOs ☐ Industry	☐ Natural Background ☒ POTWs ☐ Other:	☒ Urban Stormwater ☐ Mining
UAA Factor -131.10 (g)	2, 3, 4	☐ Undetermined	
Designated Use Assessed	☐ Aquatic Life ☐ Other	☐ Recreation	☒ Drinking Water
Parameters of Concern	☐ Bacteria ☐ Organic Hydrocarbons ☐ Temperature ☐ Chloride ☒ Other: Trihalomethanes	☐ Metals ☐ Pesticides/Herbicides ☐ Ammonia ☐ Dissolved Oxygen	☐ Nutrients ☐ Solids ☐ BOD
Status	☐ Completed & Adopted ☐ Completed & Rejected ☐ Under EPA review	☐Considered not developed ☐ Started - abandoned	☐On-going; ☒Under Consideration
Outcome	☐ Use Upgrade ☒Other: On-going	☐ Use Removal/Revision	☐ No Change in Use

Hydropower-Related UAAs for Ragged Lake and Seboomook Lake, Maine

State(s)	ME	**EPA Region(s)** 1	
Leading effort	Maine DEP		
Contacts	Dana Murch, ME DEP		
Start	2004	**End**	2006
Setting	☐ Effluent Dominated	☐ Wet-weather Impacted	☐ Urban
	☒ Other:		
Pollutant Source Category	☐ Agricultural Runoff	☐ Natural Background	☐ Urban Stormwater
	☐ CSOs	☐ POTWs	☐ Mining
	☒ Industry	☒ Other: Dam operation for electric power	
UAA Factor -131.10 (g)	3	☐ Undetermined	
Designated Use Assessed	☒ Aquatic Life	☐ Recreation	☐ Drinking Water
	☐ Other		
Parameters of Concern	☐ Bacteria	☐ Metals	☐ Nutrients
	☐ Organic Hydrocarbons	☐ Pesticides/Herbicides	☐ Solids
	☐ Temperature	☐ Ammonia	☐ BOD
	☐ Chloride	☐ Dissolved Oxygen	
	☒ Other: Water quality in lake		
Status	☒ Completed & Adopted	☐Considered not developed	☐On-going;
	☐ Completed & Rejected	☐ Started - abandoned	☐Under Consideration
	☐ Under EPA review		
Outcome	☐ Use Upgrade	☒ Use Removal/Revision	☐ No Change in Use
	☐Other:		

Central Valley Basin-Wide pH and Turbidity UAA, California

State(s)	CA	**EPA Region(s)** 9	
Leading effort	City of Roseville/Central Valley Joint Powers Association		
Contacts	Art O'Brien		
Start	2001	**End**	2007
Setting	☒ Effluent Dominated	☐ Wet-weather Impacted	☐ Urban
	☐ Other:		
Pollutant Source Category	☐ Agricultural Runoff	☐ Natural Background	☐ Urban Stormwater
	☐ CSOs	☒ POTWs	☐ Mining
	☐ Industry	☐ Other:	
UAA Factor -131.10 (g)	2	☐ Undetermined	
Designated Use Assessed	☒ Aquatic Life	☒ Recreation	☐ Drinking Water
	☐ Other		
Parameters of Concern	☐ Bacteria	☐ Metals	☐ Nutrients
	☐ Organic Hydrocarbons	☐ Pesticides/Herbicides	☐ Solids
	☐ Temperature	☐ Ammonia	☐ BOD
	☐ Chloride	☐ Dissolved Oxygen	
	☒ Other: pH, turbidity		
Status	☐ Completed & Adopted	☐Considered not developed	☒On-going;
	☐ Completed & Rejected	☐ Started - abandoned	☐Under Consideration
	☐ Under EPA review		
Outcome	☐ Use Upgrade	☐ Use Removal/Revision	☐ No Change in Use
	☒Other: Criteria revision		

Ecoregion Refinement of DO Criteria, Mermentau River Basin Example, Louisiana

State(s)	LA	**EPA Region(s)** 6	
Leading effort	Louisiana DEQ		
Contacts	Kris Pintado, LA DEQ		
Start	1998	**End**	1999
Setting	☐ Effluent Dominated	☐ Wet-weather Impacted	☐ Urban
	☒ Other:		
Pollutant Source Category	☒ Agricultural Runoff	☒ Natural Background	☐ Urban Stormwater
	☐ CSOs	☒ POTWs	☐ Mining
	☐ Industry	☐ Other:	
UAA Factor -131.10 (g)	5	☐ Undetermined	
Designated Use Assessed	☒ Aquatic Life	☐ Recreation	☐ Drinking Water
	☐ Other		
Parameters of Concern	☐ Bacteria	☐ Metals	☐ Nutrients
	☐ Organic Hydrocarbons	☐ Pesticides/Herbicides	☐ Solids
	☐ Temperature	☐ Ammonia	☐ BOD
	☐ Chloride	☒ Dissolved Oxygen	
	☐ Other		
Status	☒ Completed & Adopted	☐Considered not developed	☐On-going;
	☐ Completed & Rejected	☐ Started - abandoned	☐Under Consideration
	☐ Under EPA review		
Outcome	☐ Use Upgrade	☐ Use Removal/Revision	☐ No Change in Use
	☒Other:: Use refinement		

UAA for Ramos Forested Wetlands at Amelia, Louisiana

State(s)	LA	**EPA Region(s)** 6	
Leading effort	Louisiana DEQ		
Contacts	Kris Pintado, LA DEQ		
Start	1995	**End**	2001
Setting	☐ Effluent Dominated ☒ Other:	☐ Wet-weather Impacted	☐ Urban
Pollutant Source Category	☐ Agricultural Runoff ☐ CSOs ☐ Industry	☒ Natural Background ☒ POTWs ☐ Other:	☐ Urban Stormwater ☐ Mining
UAA Factor -131.10 (g)	1, 2, 5	☐ Undetermined	
Designated Use Assessed	☒ Aquatic Life ☐ Other	☐ Recreation	☐ Drinking Water
Parameters of Concern	☐ Bacteria ☐ Organic Hydrocarbons ☐ Temperature ☐ Chloride ☐ Other	☐ Metals ☐ Pesticides/Herbicides ☐ Ammonia ☒ Dissolved Oxygen	☐ Nutrients ☐ Solids ☐ BOD
Status	☒ Completed & Adopted ☐ Completed & Rejected ☐ Under EPA review	☐ Considered not developed ☐ Started - abandoned	☐ On-going; ☐ Under Consideration
Outcome	☐ Use Upgrade ☒ Other: Revised criteria	☐ Use Removal/Revision	☐ No Change in Use

Cuyahoga River Ship Channel UAA, Ohio

State(s)	OH	**EPA Region(s)** 5	
Leading effort	Cuyahoga Remedial Action Planning Process (RAPP)		
Contacts	Lester Stumpe		
Start	1990	**End**	1993
Setting	☐ Effluent Dominated	☐ Wet-weather Impacted	☒ Urban
	☐ Other:		
Pollutant Source Category	☐ Agricultural Runoff	☐ Natural Background	☒ Urban Stormwater
	☒ CSOs	☒ POTWs	☐ Mining
	☐ Industry	☐ Other:	
UAA Factor -131.10 (g)	3	☐ Undetermined	
Designated Use Assessed	☒ Aquatic Life	☐ Recreation	☐ Drinking Water
	☐ Other		
Parameters of Concern	☐ Bacteria	☐ Metals	☐ Nutrients
	☐ Organic Hydrocarbons	☐ Pesticides/Herbicides	☐ Solids
	☐ Temperature	☐ Ammonia	☐ BOD
	☐ Chloride	☒ Dissolved Oxygen	
	☐ Other		
Status	☒ Completed & Adopted	☐ Considered not developed	☐ On-going;
	☐ Completed & Rejected	☐ Started - abandoned	☐ Under Consideration
	☐ Under EPA review		
Outcome	☐ Use Upgrade	☐ Use Removal/Revision	☐ No Change in Use
	☒ Other: New use assignment with site-specific criteria– no previous use		

APPENDIX C

USE ATTAINABILITY ANALYSIS RESOURCES

This appendix contains a list of available resources that may be beneficial to those considering a UAA or currently involved in one. The resources are organized into federal, state, and other categories. Internet links that were functional at the time this report was published are also provided.

FEDERAL

Designation of Uses; EPA. Chapter 2 from the Water Quality Standards Handbook; Second Edition. EPA 823-B-95-002, EPA, Office of Water, 1995.
http://www.epa.gov/waterscience/standards/handbook/handbookch2.pdf

Draft Region VIII Guidance for Use Attainability Analyses for Aquatic Life Uses; EPA. 1995.
http://www.epa.gov/Region8/water/wqs/UAAGUIDE.pdf

EPA Region VIII Recreation Use Attainability Analysis Worksheet; Vodehnal, D.; EPA Letter addressed to State or Tribal Colleague, April 8, 1994.
http://www.epa.gov/Region8/water/wqs/Worksheet94wCoverLtr.pdf

Integrating Ecological Risk Assessment and Economic Analysis in Watersheds: A Conceptual Approach and Three Case Studies; EPA/600/R-03/140R, EPA. Federal Register Vol. 69, No. 26, 2003.

Interim Economic Guidance for Water Quality Standards Workbook; Appendix M to the "Water Quality Standards Handbook; Second Edition." EPA 823-B-95-002, EPA, Office of Water, 1995.

Plan for Supporting States and Tribes on Designated Use Issues; EPA, August 9, 2004.
http://www.deq.state.id.us/water/assist_business/workshops/national_use_designation_plan_uaa_workshop_handout.pdf

Proceedings Summary Report, Great Lakes Environmental Center. The National Symposium Designating Attainable Uses for the Nation's Waters. Marriott Wardman Park Hotel, Washington, DC, 2002.

Regional and National Overview of Use Attainability Analysis; Kubo, T.; EPA Region 10, Oregon Operations Office: Portland, OR, September 20, 2004.
http://www.deq.state.or.us/WQ/standards/UAA/Meetings/UAAW092004EPAOverview.pdf

State Water Quality Standards Approvals Use Attainability Analysis Submittals. EPA. Internal EPA Memo to Patrick W. Tobin and David A. Fierra, November 28, 1984.
http://www.epa.gov/waterscience/library/wqstandards/uaasubmittals.pdf

Technical Support Manual: Waterbody Surveys and Assessments for Conducting Use Attainability Analyses-Volume 2: Estuarine Systems. EPA440486038. EPA Office of Water Regulations and Standards: Washington, DC. January 1983.

Technical Support Manual: Waterbody Surveys and Assessments for Conducting Use Attainability Analyses. EPA440486037. EPA Office of Water Regulations and Standards: Washington, DC. November 1983.

Use Attainability Analyses (UAAs); EPA, Includes Introduction to UAAs and Nationwide Case Studies. March, 2006. http://www.epa.gov/waterscience/standards/uaa/index.htm

Water Quality Criteria; Chapter 3 from the "Water Quality Standards Handbook; 2nd ed." EPA 823-B-95-002. Office of Water, 1995. http://www.epa.gov/waterscience/standards/handbook/handbookch3.pdf

STATE

Aquatic Life - Use Attainability Studies for Flowing and Impounded Water Bodies; Pennsylvania Department of Environmental Protection, June 1990.

Combined Sewer Overflow (CSO) Long-Term Control Plan Use Attainability Analysis Guidance; Water-003-NRD, Indiana Department of Environmental Management, 2001. http://www.in.gov/idem/rules/policies/water/003.pdf

Compilation of Existing Guidance for the Development of Site-Specific Water Quality Objectives in the State of California; California Environmental Protection Agency, June 2003. http://www.swrcb.ca.gov/docs/sitespec/sitespecificwqo_cover.pdf

Draft State of California S.B. 469 TMDL Guidance A Process for Addressing Impaired Waters in California Appendix C, Use Attainability Analysis and Site Specific Objectives; California State Water Resources Control Board, March 2005. http://www.swrcb.ca.gov/docs/sitespec/sitespecificwqo_appxc.pdf

Draft Use Attainability Variance Guidance; Virginia Department of Environmental Quality, July 6, 2001.

Guidance Document for Use Attainability Analyses (UAAs); Kansas Department of Health and Environment, Bureau of Environmental Field Services: Topeka, KS, December 1, 2001. http://www.kdheks.gov/befs/uaas/UAAGuidance.pdf

Montana Water Quality Act Statute 75-5-312. Temporary Water Quality Standards; MT Department of Environmental Quality, 2003. http://www.deq.mt.gov/wqinfo/Laws/WQA2003.pdf

Recreational Use Attainability Analysis Protocol; Missouri Department of Natural Resources, November 3, 2004. http://www.dnr.mo.gov/env/wpp/wqstandards/uaa/wpp_wqs_uaa.pdf

Recreational Use Classification Guidance; Version 1.1, Colorado Department of Public Health and Environment, January 2003. http://www.cdphe.state.co.us/wq/Assessment/Assess_pdf/RecUAAGuidev11.pdf

Unified Protocols for Beneficial Use Assignment for Oklahoma Wadable Streams (Use Attainability Analysis); Technical Report TRWQ2001-1, OklahomaWater Resources Board, 2001. http://www.owrb.state.ok.us/studies/reports/reports_pdf/TR2001_1_Protocols_Wadable.pdf

Use Attainability Analysis Guidance for Washington State; Draft – Version 1.2, Washington Department of Ecology, July 2005. http://www.ecy.wa.gov/programs/wq/swqs/uaa_docs/draft_uaa-guidance-072005_v1.2.pdf

Water Body Assessment Guidance; 2nd ed – final, Idaho Department of Environmental Quality, January 2002. http://www.deq.state.id.us/water/data_reports/surface_water/monitoring/publications.cfm#wbag

OTHER

Cracking the UAA Code: Wet Weather Water Quality Assessments; National Association of Clean Water Agencies (NACWA) Conference. Chicago, IL, October 2004. http://www.nacwa.org/meetings/04UAA/

Exploring Use Attainability Analysis; National Association of Clean Water Agencies (NACWA), Water Environment Research Foundation (WERF), WERF 2005 Web Seminar Series, August 17, 2005.

Use Attainability Analysis: A Comprehensive UAA Technical Reference; Water Environment Research Foundation (WERF), Project 91-NPS-1, 1997.

Collaborative Water Quality Solutions: Exploring Use Attainability Analyses; National Association of Clean Water Agencies (NACWA), Water Environment Research Foundation (WERF), Project 04-WEM-7, 2005.

IWA Publishing's authorised EU representative for General Product Safety Regulations is Diane D'Arras, 15 rue Duret, 75116 Paris, France, e-mail: safety@iwap.co.uk.

Printed and bound by CPI Group (UK) Ltd, Croydon, CR0 4YY
17/09/2026
02218811-0018